T0241408

SpringerBriefs in Environmental Science

SpringerBriefs in Environmental Science present concise summaries of cutting-edge research and practical applications across a wide spectrum of environmental fields, with fast turnaround time to publication. Featuring compact volumes of 50 to 125 pages, the series covers a range of content from professional to academic. Monographs of new material are considered for the SpringerBriefs in Environmental Science series.

Typical topics might include: a timely report of state-of-the-art analytical techniques, a bridge between new research results, as published in journal articles and a contextual literature review, a snapshot of a hot or emerging topic, an in-depth case study or technical example, a presentation of core concepts that students must understand in order to make independent contributions, best practices or protocols to be followed, a series of short case studies/debates highlighting a specific angle.

SpringerBriefs in Environmental Science allow authors to present their ideas and readers to absorb them with minimal time investment. Both solicited and unsolicited manuscripts are considered for publication.

Kodoth Prabhakaran Nair

Extractive Farming or Bio Farming?

Making a Better Choice for the 21st Century

Kodoth Prabhakaran Nair
Formerly Professor
National Science Foundation
The Royal Society, Belgium & Senior Fellow
Alexander von Humboldt Research Foundation
Bonn, Germany

ISSN 2191-5547   ISSN 2191-5555   (electronic)
SpringerBriefs in Environmental Science
ISBN 978-3-031-34694-1   ISBN 978-3-031-34695-8   (eBook)
https://doi.org/10.1007/978-3-031-34695-8

This Springer imprint is published by the registered company Springer Nature Switzerland AG
The registered company address is: Gewerbestrasse 11, 6330 Cham, Switzerland

India's great President, late Dr A.P.J. Abdul Kalam launching the book "Issues in National and International Agriculture", authored by Professor Kodoth Prabhakaran Nair, in Raj Bhavan (Office of the Governor of the State), in Chennai, Tamil Nadu, India

I dedicate this book to my wife, Pankajam, a nematologist trained in Europe who made the selfless decision to give up her profession over four decades ago to become a homemaker when we had our son, Kannan, who is now a doctor, and our daughter, Sreedevi, who is now an engineer. Pankajam is my everything and sustains me on this difficult journey we call life.

Acknowledgments

I greatly appreciate the efficient support of Ms. Margaret Deignan, Senior Editor, Springer.

Ms. Fairle T Thattil, Project Coordinator (Books), and her team also did a marvelous job in bringing out this remarkable book.

Contents

Chapter 1
Introduction

Abstract The chapter discusses, at length, the environmental hazards of the highly chemical-centric farming, euphemistically known as the "green revolution" in South Asia, in particular, in India. It discusses the potentials of bio farming as an effective alternative to the highly extractive farming centered on chemical fertilizers, to contain the huge environmental hazards, in particular soil-related hazards.

Keywords Green revolution · Soil chemical fertilizer · Microbial fertilizer

Global farming is at a crossroads. When man, who was a hunter and gatherer, more than 9000 years ago, shifted his attention from hunting and gathering to stationary agriculture, which subsequently came to be known as the "agricultural revolution," soil has been at the center of his focus. The highly chemical-centric and extractive farming, euphemistically carrying the name "green revolution," is an extreme extension of this shift in focus, which is a shift in focus from the original agricultural revolution, which was nothing but tending to plants rather than hunting and gathering their fruits. In the process, it is the soil that paid a huge environmental price. Much land has degraded and become unsuitable for agriculture since a century ago. The 1992–1993 World Resources Report from the United Nations contains very alarming conclusions. For example, nearly ten million hectares of best farmlands of the world have been so much ruined by human activity, of which one is the green revolution. Over 1.2 billion hectares in the world that have been very seriously damaged can only be restored to normalcy at a great cost. The loss in soil capability could mean that there will be enormous food shortages in the next two to three decades. A serious consequence of this is, the people of the disadvantaged nations will suffer the most as usual. Two-thirds of the seriously eroded land is in Asia and Africa. About 25% of the cropped land in Central America is moderately to severely damaged. In North America, the percentage is small, only about 4.4%. Since the green revolution, food production has declined dramatically in 80 developing countries during the past decade. Soil degradation is the major cause. Nearly 40% of world's farming is done on small parcels of land, 1 hectare or less. Ignorance and poverty characterize the situation. In India, alone, of the 328.73 million hectares of geographical area, more than 120.40 million hectares have now degraded soils, thanks to the green revolution. The state of Punjab, the "cradle of Indian green revolution" is the living example of this environmental hazard, with specific reference to soil resources. There are thousands of hectares where not even a blade of grass will grow without huge investments in soil reclamation. And this means a huge drain of national resources. All of this proves beyond a shadow of doubt the centrality of soil in human sustenance.

The book *Silent Spring*, by Rachel Carson, launched in 1962, spurred revolutionary changes in the laws relating to air, land, and water. Rachel Carson's passionate concern for the future of the planet reverberated powerfully throughout the world and was instrumental in launching the "environmental movement." Indeed, the intelligent management of global soil resources should be the bedrock of this environmental movement. This book will examine the critical question whether or not, to meet the growing demands of a burgeoning global population, are there other viable alternatives to the highly chemical-centric and extractive farming.

The Concept of Soil: What Does the Word "Soil" Imply?

1.1 The Challenge

During the World Soil Science Congress, held in Hamburg, the Federal Republic of Germany, in 1986, I referred to soil as "soul of infinite life," in the plenary session, to the mirth of the assembled audience. The intention was to substitute the words, "soul" (for the letter s), "of" (for the letter o), "life" (for the letter l). Most in the assembled audience thought that soil was an "inert" matter; hence, what life and what soul? But if one critically examines what goes round on the planet, one can appreciate the intensity and meaning of this phrase. In fact, it is soil that gives the very life and meaning to life on earth – be it humankind, animal, or plant. When soils are ruined, civilizations fall. The Roman Empire collapsed when its North African soils desertified. The same is true of the Thar Desert in India, which once was a dense jungle. It is the mindless exploitation of soil by man that leads to his ruination.

Soils are regarded by the International Policy Community as increasingly important in world development issues, such as food security, poverty alleviation, land degradation, and the provision of environmental services (Wood et al. 2000). Soils are a crucial component of terrestrial ecosystems and a determinant of their capacity to produce goods and services. Soils exert production, buffering, filtering, and biological functions. Solar energy, carbon dioxide and nitrogen from the air, and nutrients from the soil are converted into plant products that provide animals and humans with food, fiber, and biofuels. Soils hold water from episodic rainfall or irrigation as well as nutrients applied as organic inputs or mineral fertilizers, releasing them at rates plants can utilize for longer periods of time. Soil biota decomposes organic minerals, cycle nutrients, and regulates gas fluxes to and from the atmosphere. Soils filter nonhazardous and toxic compounds through surface adsorption and precipitation reactions and largely determine the quality of terrestrial waters. Soils, therefore, deliver many of our basic needs and play a central role in determining the quality of the environment.

Though we all, as soil scientists, will not dispute what has been written above, how is it that soil science has been pushed to the backyard, while others have made spectacular advances or, at least, others are noticed by people, at large, the world over, while we continue to be ignored by the world? In fact, the most classic example, of late, is the science of genetically modified organisms (GMOs), be they of plant or animal origin. Notwithstanding the controversies surrounding them, they have captured the attention of people all over the world. Ironically, none stops to think that, after all, even a genetically modified plant, say, for instance, a Bt cotton plant, cannot grow in outer space but needs a fertile soil to grow on. Tragically, the science of soil is not even recognized as a "science" in the sense others are, as, for example, plant science or medicine, or even economics. More often than not, my wife, a trained nematologist (trained in Europe), and two grown-up children, an elder son who is a doctor and a younger daughter who is an engineer, chide me as to where I have reached in life during the last more than five decades of professional commitments, dirtying my hands with soil and thinking of the science of soil as a "science." Their difficulty in trying to understand what I have done, and continue to do, is what public awareness people call a "brand failure." Our lack of visibility is related to our culture as "reductionist scientists," operating largely within our limited scientific circles and with land users who utilize our knowledge of soil science and our extension services. The take-home message is that we soil scientists are the problem. But we can become the solution by undertaking the kinds of synthesis research that are of direct use to policy makers and communicating to them in a way they readily accept it and put it to use. A classic example of global success is that of the Bt technology, where policy makers, starting from the United States of America to India (where the science of the Bt technology was totally unknown even up to as late as 2002), got the total involvement of the governments, and so the policy makers who run them. Soil science has been brilliantly informed by reductionist physics and chemistry, poorly informed by biology, ecology, and geography and largely uninformed by the social sciences (Swift 1999). In a survey of the global environment, The Economist magazine (6 July 2002) reported that leading experts could not reach a firm conclusion about the state of the environment because much of the information they needed was incomplete or missing altogether. This article went on to say that businessmen always say "What matters gets measured." While soil scientists cannot be accused of not measuring soil properties, we are perhaps guilty in the lack of synthesis, integration, and interpretation of those measurements as they relate to environmental goods and services. We should certainly take on this challenge. It is in this context that a revolutionary soil management technique like "The Nutrient Buffer Power Concept," developed by this author, over more than three decades of research in Europe, Africa, and Asia, could be made to be an acceptable idea for the policy makers so that it attains the status of a "brand."

1.2 Soils and Sustainable Agriculture

During the nineteenth century and the first half of the twentieth century, when the global population was only 38% of its current level in 2006, the primary objective of soil management was to ensure agronomic productivity to meet the demands of a growing world population of roughly 2–3 billion. However, the demands on soil resources today are vastly different in a densely populated and rapidly industrializing world. For instance, India has now surpassed China as the most populous country in the world, with a population of over 1.3 billion people.

Despite the initial success of the highly chemical-centric farming practices known as the "green revolution," which led to a significant increase in food production, the situation in India, particularly in the state of Punjab, the birthplace of the Indian green revolution, has deteriorated. There are vast areas of land where not even a blade of grass can grow without costly soil reclamation. Out of India's total geographical area of 328.73 million hectares, over 120.40 million hectares now suffer from degraded soils, largely due to the unsustainable practices of the green revolution. Punjab serves as a prime example, where the excessive use of urea to support the cultivation of "miracle" rice and wheat varieties has severely depleted soil resources. Groundwater is polluted, water sources are no longer potable, and the indiscriminate use of pesticides and herbicides has led to an increase in cancer cases. Gurdaspur district in Punjab has even earned the dubious title of being the "cancer capital" of India.

Furthermore, the chemically centric green revolution has contributed to global warming. The excessive use of synthetic nitrogen fertilizers, such as urea, to boost food production is a primary driver for the atmospheric concentration of nitrous oxide. Nitrous oxide, released during urea hydrolysis, escapes into the stratosphere, where it traps radiant heat and contributes to global warming. Nitrous oxide molecules remain in the stratosphere or atmosphere for an average of 114 years before being removed by natural sinks or destroyed through chemical reactions. The warming impact of one pound of nitrous oxide is almost 300 times greater than that of one pound of carbon dioxide. These issues are discussed in detail by Nair (2019) in his book *Combating Global Warming: The Role of Crop Wild Relatives for Food Security* (Springer Nature Switzerland AG 2019). According to the author, the green revolution has contributed as much as 35% to the current global warming, particularly in countries like India.

All of these concerns emphasize the critical role of soil in these complex challenges. The question of how to manage soils for sustainable crop production is of utmost importance.

1.3 Ensuring and Advancing Food Security

During the biblical era, the global population was approximately 0.2 billion, which increased to 0.31 billion over the next 1000 years by 1000 AD. Subsequently, over the following 1000 years, the global population experienced a 20-fold increase to 6 billion by 2000 AD. According to projections by Cohen (2003), the global population is expected to reach 9.4 billion by 2050. Strikingly, all the projected population growth is anticipated to occur in the developing world, with an estimated increase of 3.5 billion people. This raises the critical question of how humanity will feed the additional mouths, especially considering the soil degradation issues prevalent in developing countries like India, as discussed earlier.

Nutrient depletion and imbalances in soil pose significant challenges for crop growth and final yield. These concerns are of global importance to soil scientists. Despite the fact that more than 50% of a plant's yield is determined by nutrient availability, it is also the aspect that is least resilient to management. Therefore, understanding the dynamics of soil nutrient bioavailability and implementing intelligent management strategies become paramount for sustainable agriculture and soil management practices.

Tan et al. (2005) estimated that the global rates of nutrient depletion will be as follows:

Nitrogen: 18.7 kg ha^{-1} year^{-1}
Phosphorus: 5.1 kg ha^{-1} year^{-1}
Potassium: 38.8 kg ha^{-1} year^{-1}

The above rates were estimated for 59%, 85%, and 90% of the harvested area in 2000. Tan and his colleagues further estimated the global annual nutrient deficit as:

Nitrogen: 5.5 Tg
Phosphorus: 2.3 Tg
Potassium: 12.2 Tg

1.4   What do the Above Data Prove?

The deficits mentioned above lead to a total loss of 1.136 million tons of food grains globally. Soil nutrient depletion can be attributed to several factors, including:

1. Lack of or insufficient use of fertilizers
2. Unbalanced use of fertilizers
3. Losses caused by soil erosion

Among these factors, the second point holds significant importance and warrants further discussion. It is crucial to evaluate the effectiveness of strategies that promote and maintain positive nutrient and carbon budgets in managed ecosystems.

Nutrient depletion occurs when the removal of nutrients (through harvest, erosion, leaching, and volatilization) exceeds nutrient input. One notable example is the emission of nitrous oxide resulting from urea hydrolysis, which was mentioned earlier in this chapter. It is important to emphasize that nutrient depletion, driven by extractive farming practices, has had a severe negative impact on farming and crop yields in regions like South Asia (SA) and sub-Saharan Africa (SSA) (IFDC 2006).

The consequences of extractive farming, commonly known as the "green revolution," serve as a vivid illustration of these detrimental effects.

The need for "carbon farming".

Carbon sequestration in terrestrial ecosystems, including soils, trees, wetlands, and other natural habitats, as well as improving soil quality to promote the sequestration of CH_4 and reduce N_2O emissions, is a critical concern that requires the expertise of various scientific disciplines. Soil scientists, crop scientists, agronomists, foresters (both researchers and managers), and wetland ecologists all play important roles in addressing this issue. Nair (2019) has provided compelling arguments on this topic, emphasizing the potential of "Crop Wild Relatives," a largely untapped resource, in achieving these goals. Their exploration and utilization could greatly contribute to carbon sequestration and mitigation of greenhouse gas emissions in agricultural and natural systems.

What are the basic principles of sustainable soil management?

The following 10 principles of sustainable soil management must be considered:

Principle 1
Soil resources are not uniformly distributed across biomass and geographic regions. Highly productive soils are often found in densely populated areas, such as India, and have already been converted into managed ecosystems like croplands, grazing lands, forests, and energy plantations.

Principle 2
Most soils on the planet are susceptible to degradation due to misuse or mismanagement. Anthropogenic soil degradation is a result of desperate situations and the helplessness of resource-poor/marginal farmers and small landholders. This is particularly evident in India, where human greed, short-sightedness, and a lack of proper planning have led to quick but ultimately worthless results in the long run.

Principle 3
Decline in soil fertility, quality, and accelerated soil erosion are often caused by improper crop management practices rather than the choice of crops themselves. Restorative soil and water management practices are essential for realizing the full potential of farming systems. Indiscriminate and excessive tillage, irrigation, and fertilizer use can contribute to soil degradation, as exemplified by the chemically centric "green revolution."

Principle 4
Increases in mean annual temperature and decreases in mean annual precipitation can accelerate soil degradation. Soils in hot and arid regions are more prone to degradation and desertification, but even arctic climates can experience desertification due to soil mismanagement.

Principle 5
Soil can either be a source or sink of greenhouse gases, such as methane (CH_4), carbon dioxide (CO_2), and nitrous oxide (N_2O), depending on land use and soil management. Soil is a sink of atmospheric CO_2 under those land use and

management systems that create a positive carbon (C) budget and gains exceed losses. Soil is a source of atmospheric CO_2 when the ecosystem C budget is negative and losses exceed the gains. Soils are a source of radiatively active gases with extractive farming (read "Green revolution"), which create a negative nutrient budget that degrades soil quality (as has widely taken place in Punjab state, India, consequent to "green revolution). And a sink with restorative land use and judicious soil management practices, especially fertilizer management, which create positive C and nutrient budgets and conserve soil and water while improving soil structure. This is where the relevance of "The Nutrient Buffer Power Concept" comes into great scientific relevance.

Principle 6

The most important fact to remember in farming is that soils are a nonrenewable asset over a human time frame of decadal and generational scales, which are nonrenewable, on a geological scale, conforming to a centennial/millennial time frame. With burgeoning world population, which scientists project at 10 billion by 2100 AD, restoration of degraded and desertified soils over a centennial/millennial time frame is next to an impossibility. The heavy human demands for food on this nonrenewable, and if I may add, an invaluable global asset, such as soil, it becomes extremely imperative that our approach to soil management must be visionary, not with short-term gains but with long-term benefits for all life – human, animal, and plant.

Principle 7

Both natural and anthropogenic perturbations are a test for the resilience of the soil. This, of course, depends on the physical, chemical, and biological processes that take place in the soil substrate. Benevolent physical and chemical processes enhance soil's resilience. And this happens only under favorable soil physical conditions, such as optimum/good soil structure, tilth, aeration, water retention, transmission, and edaphological conditions, such as soil temperature.

Principle 8

Soil organic matter restoration rate of the soil organic pool is extremely slow. Its depletion, more often than not, is so very rapid. Global deforestation for human greed is the best example.

Principle 9

Both an architectural design and soil structure are similar to each other, inasmuch as functional aspects conform to stability, and continuity of macro-, meso-, and micropores, which are the sites of physical, chemical, and biological processes, which support soil's life supportive functions. Sustainable soil management systems, which are site-specific, enhance greatly soil stability and continuity of pores, specified above, and voids, over a time frame, also, under diverse land use systems.

It is, thus, the superb ingenuity and foresight of the "soil manager" – be it a soil scientist/crop scientist/biologist – that ensures the premier dictum of this author of soil as a "soul of infinite life," not merely to be handled as an "inert" material, at will, for personal gain and profit.

Principle 10
Ensuring and enhancing trend in net primary productivity per unit of off-farm inputs, such as fertilizers and water, along with improvement in soil quality and ancillary ecosystem services, such as enhancement of carbon pool, and improvement in quantity and quality of freshwater and renewable water resources, with concomitant increase of biodiversity, must be the prime focus in soil management.

As said in the initial section of this chapter, soil is an invaluable asset and the "soul of infinite life." It is imperative that the current generation hands over the global soil resources, to the generation next, in as good a state as possible, as was inherited, from the past generation. If we fail, the fate of humankind will be similar to that of civilizations that lost their perspective, like the Mayan, Incas, Indus, and Mesopotamian.

1.5   What Does the Concept of Soil Health Mean?

Soil health is the continued capacity of soil to function as a vital living ecosystem that sustains plants, animals, and humans and connects agricultural and soil science to policy, stakeholder needs, and sustainable supply chain management. Historically, soil assessments focused on crop production, but today, soil health also includes the role of soil in water quality, climate change, and human health. However, quantifying soil health is still dominated by chemical indicators, despite growing appreciation of the importance of soil biodiversity, owing to limited functional knowledge and lack of effective methods. In this perspective, the definition and history of soil health are described and compared with other soil concepts. There are many ecosystem services provided by soils, the indicators used to measure soil functionality and their integration into informative soil health indices. Scientists should embrace soil health as an overarching principle that contributes to sustainability goals, rather than only a property to measure.

1.6   A Looming Global Food Crisis

Though the Indian meteorological department has been reassuring Indians that there would be a regular monsoon, all the ambient factors indicate otherwise. April temperatures are soaring. In India's wheat granary, Punjab, the ambient temperature is 41 degrees Celsius (daytime) with night temperature at 23 degrees Celsius, which will lead to a huge grain loss due to shrinking and shattering of grains. This is the harvest time. The other parts of India are no better.

1.7   The Global Scenario

Is the world headed for another year of food crisis? All the available indicators are pointing toward it. If this happens, which is quite a likelihood, 2023 will be the third consecutive year with a record price rise of food, especially of cereal grains. During the first 22 years of this century, the world has already dealt with three major global food price hikes, in 2007–2008, 2010–2011, and 2021–2022. This year will add to this record-breaking period of high global food prices.

According to the latest World Bank's Food Security Update, around four-fifths of low-income countries and more than 90% of lower middle-income countries have witnessed year-on-year food price escalation in excess of 5%. The "cost of living crisis," caused by skyrocketing food prices, is currently gripping the world to such an extent that the "Global Risk Report 2023" by the World Economic Forum (WEF) has found this as the topmost severe threat over the coming next 2 years.

1.8   What is the Situation in India?

Due to a gradual switchover to the cultivation of commercial crops, the area under the cultivation of food grains has steadily decreased in the last 5 years by about 30%. As a consequence of increasing incomes, the consumption patterns of the people have undergone a significant change.

For almost 2022, more than 20% rise in food inflation since May occurred during the last 3 months, witnessing 30% inflation. Data from the wholesale price index (WPI) has clearly indicated that cereals and other food items have been on a rising trend since 2021.

The Union Ministry of Finance has already warned that food prices will rise in 2023 due to various reasons, with extreme weather events, and the looming El Nino being the prime factors, which would impact overall harvests. The World Food Programme (WFP) has forecast that 345.2 million persons will be "food insecure" in 2023. This is more than double the number in 2020, when the COVID-19 pandemic started. It also means that 200 million more persons are food insecure in comparison to pre-pandemic levels.

1.9   What is the Future?

It looks as though the last century's achievement in curbing hunger will be undone in the current century. In the last century, the world had nearly seen an eradication of famines and witnessed the rise of a system that enabled aversion of extreme food scarcity, through massive relief operations. The rise of democratic institutions also helped with effective responses to such dire situations. The last century also reported a drop in severe conflicts that traditionally triggered famines. But that situation is now being replaced by pockets of global conflicts. The Russian-Ukranian war, of the last almost 2 years now, is a typical example. Its adverse impact on fossil fuel and food supply is clearly visible. Sadly, there is a looming crisis between China and Taiwan now.

The world is witnessing famines or famine-like situations, notwithstanding the developments of the past. Extreme weather events and climatic factors are replacing international wars and conflicts as the main reasons for creating famines or famine-like situations. Russian invasion of Ukraine caused the food prices to escalate in 2021–2022.The number of climate events damaging crops and displacing people also played a definitive role in escalating food prices and people's affordability for food during the last year. Unprecedented drought spell caused the famine of Horn of Africa. According to WFP, 0.9 million persons are already surviving in famine-like conditions right now. This population segment has escalated by 10 times during the last 5 years, pushing it to desperate survival situations.

As per the projections of the International Food Policy Research Institute in Washington (IFPRI), on average "a 5 percent increase in the real price of food increases the risk of wasting by 9 percent and severe wasting by 14 percent." This, undoubtedly, will further aggravate the already existing burden of malnutrition, quite possibly, leading to mortality. This is a typical famine situation.

In the past, while the mortality rate from a famine was quite high and it could be reduced to near zero, the current century might see a reversal of this pattern. The most distinguishing fact of this is that in the current century, the crop damage and the concomitant price hike in food items, especially food grains, are the results of climate events. And, sadly, the victims are nearly the same, namely, the vulnerable populations of the poorest and low-income developing countries.

The above grim scenario is a direct consequence of the global warming effect lucidly described by Nair (2019), contributed largely by a highly soil-extractive farming.

References

Cohen JE (2003) The human population: next half century. Science 320:1172–1175
IFDC (2006) African soil exhaustion. Science 312:31
Nair KPP (2019) Combating global warming – the role of crop wild relatives for food security. Springer Nature Switzerland AG
Swift MJ (1999) Integrating soils, systems and society. Nature & Resources 35:12–20
Tan ZX, Lal R, Wiebe KD (2005) Global soil nutrient depletion and yield reduction. J Sustain Agric 26:123
Wood S, Sebastian K, Scherr SJ (2000) Pilot analysis of global ecosystems: agroecosystems. International Food Policy Research Institute/World Soil Resources Institute, Washington, DC

Chapter 2
What are Microbial Fertilizers and What is Their Role in Bio Farming?

Abstract The chapter provides a comprehensive analysis of different types of microbial fertilizers, their benefits, and their significance in bio farming.

Keywords Microbial bio fertilizer · Bio farming

There are several types of microbial fertilizers available, including nitrogenous and phosphatic options such as phosphate-solubilizing microbial fertilizer as well as potash-solubilizing bacteria (KSB), zinc- and silica-solubilizing bacteria (ZSB), and plant growth-promoting rhizobacteria (PGPR). Some of the most significant options are discussed below.

2.1 Nitrogenous Microbial Fertilizers and Their Role in Bio Farming

Nitrogen is a crucial plant nutrient, alongside phosphorus and potassium (NPK), required in significant quantities for optimal crop production. Although nitrogen is abundant in the atmosphere (78–80%), most crop plants like rice, wheat, and maize cannot utilize it in this gaseous state. However, legumes such as field beans (*Vigna unguiculata*) and red gram (*Cicer arietinum*) can fix atmospheric nitrogen through symbiosis with a microbe called *Rhizobium*. The term *Rhizobium* originates from the Greek word *rhiza* meaning root and *bios* meaning life. The scientist Frank was the first to name it *Rhizobium,* which when translated into Latin means "living in roots." Some of the *Rhizobium* species were subsequently reclassified based on the phylogenetic analysis. Currently, there are 49 rhizobial species and eleven non-rhizobial species. The taxonomic position of *Rhizobium* can be explained as follows:

Kingdom: Bacteria
Phylum: Proteobacteria
Class: Alpha-proteobacteria
Order: Rhizobiales
Family: Rhizobiaceae
Type species: *Rhizobium leguminosarum* and *Rhizobium lentil*

2.2   The Morphology of Rhizobium

Rhizobium, a gram-negative motile bacterium, establishes a unique symbiotic relationship with various leguminous plants, including cowpea (*Vigna unguiculata*), pea (*Pisum sativum*), soybean (*Glycine max*), and alfalfa (*Medicago sativa*). This symbiosis leads to the formation of specialized structures called root nodules when Rhizobium roots associate with a host plant. Through a process known as symbiotic nitrogen fixation, these bacteria convert atmospheric nitrogen into ammonia. The resulting ammonia serves as a vital source of nitrogen for the host plant, which is the primary essential nutrient for all crops, excluding the leguminous varieties mentioned above. In addition to *Rhizobium*, other genera such as *Azorhizobium* and *Bradyrhizobium* can also nodulate and contribute to atmospheric nitrogen fixation.

Given that nitrogen is a crucial element for crop production, the utilization of fossil fuel-based nitrogen suppliers, such as chemical and synthetic urea, can prove to be prohibitively expensive for impoverished or marginalized farmers in developing countries like India. In this context, adopting a leguminous crop as part of a rotation cycle, such as wheat/alfalfa/cowpea rotation, would offer a significantly more advantageous alternative compared to the traditional rice/wheat rotation.

Rhizobium sp. is characterized as a fast-growing bacterium, whereas *Bradyrhizobium* exhibits slower growth. Rhizobia, for the most part, demonstrate a weak affinity for absorbing Congo red dye (diphenyldiazo-bis-a-naphthylamine sulfonate). The following list highlights some of the frequently encountered species of Rhizobium:

1. *Rhizobium leguminosarum*
2. *R. alamii*
3. *R. lentils*
4. *R. japonicum*
5. *R. metallidurans*
6. *R. smilacinae*
7. *R. phaseoli*
8. *R. trifolii*

Rhizobium species can be categorized into different groups based on their growth rates. It is important to note that not all *Rhizobium* species are capable of nodulating. To facilitate classification and study, *Rhizobium* species were assigned into cross-inoculation groups. The following are some examples of cross-inoculation groups:

A. Alfalfa group: *Rhizobium meliloti* which infects and nodulates roots of *Medicago sativa* and *Melilotus officinalis* (sweet yellow clover)
B. Bean group: *Rhizobium phaseoli* which infects and nodulates roots of genus *Phaseolus vulgaris* (common bean)
C. Clover group: *Rhizobium trifolii* which infects and nodulates roots of genus white clover (*Trifolium repens*)
D. Cowpea group: *Rhizobium* sp. which infects and nodulates roots of cowpea (*Vigna unguiculata*), Pigeon pea (*Cajanus cajan*) and groundnut (*Arachis hypogaea*)
E. Lupine group: *Rhizobium lupine* which infects and nodulates roots of lupines and serradella (Ornithopus, "bird's-foot, *O. compressus* (yellow serradella), *O. micranthus, O. pinnatus, O. perpusillus* and *O. sativus*))
F. Soybean group: *Rhizobium japonicum* which infects and nodulates roots of soybean (*Glycine max*)

2.3   The Mode of Infection

Although *Rhizobium* species can be found in most soils, their natural habitat is within leguminous plants such as soybean, red gram, or cowpea. As a result of a symbiotic association between the bacterium and host plant, nodules are formed on the plant's roots, which contain abundant fixed nitrogen from the atmosphere. Flavonoids, which are released by the host leguminous plant, play an essential role in attracting *Rhizobium* species and facilitating their absorption through the bacterial membrane. These flavonoids also activate nod genes, which are involved in producing nod factors. Once the symbiotic relationship is established, transcription of nod genes occurs.

2.4   The Process of Nodulation

Upon contact with the root hair of a specific host plant, such as soybean, the bacterium *Rhizobium* initiates a series of interactions. The plasma membrane of the plant cells undergoes invagination, facilitating the penetration of Rhizobium into the cell. Subsequently, the host plant generates new cell wall material, enabling the bacterium to penetrate deeper into the root hairs. This symbiotic process leads to the formation of two distinct types of nodules outlined below:

A. *Determinate nodules*

The nodules formed as a result of this symbiotic relationship exhibit a spherical shape and possess prominent lenticels. These specific nodules are predominantly found in host plants such as soybean (*Glycine max*) and are commonly observed in tropical and sub-tropical regions around the world. However, they are typically absent in temperate regions. Notably, these nodules are known to produce ureide products as part of their physiological processes.

B. *Indeterminate nodules*

In contrast to the spherical determinate nodules, the indeterminate nodules have a cylindrical shape and are frequently branched. They are commonly found in association with plants such as alfalfa (*Medicago sativa*), sweet yellow clover (*Melilotus officinalis*), and pea (*Pisum sativum*). These indeterminate nodules are predominantly observed in temperate regions worldwide. Unlike their counterparts, they are known to produce amide products as part of their metabolic processes.

2.5   The Process of Nitrogen Fixation

In mature nodules, nitrogen is fixed from the atmosphere through a process that involves the conversion of gaseous nitrogen (which constitutes approximately 78–80% of the atmosphere) into ammonia. The nif genes in Rhizobium bacteria encode the components of the nitrogenase enzyme complex, which includes the structural subunit of dinitrogenase reductase and the two subunits of dinitrogenase, encoded by the nifH, nifD, and nifK genes, respectively. The following factors are essential for nitrogen fixation to occur:

1. Atmospheric nitrogen
2. ATP energy
3. Enzymes (dinitrogenase and dinitrogenase reductase)

During the process of nitrogen fixation, the nitrogenase enzyme is responsible for breaking the bonds that hold nitrogen atoms to other atoms, which requires a significant amount of energy. *Rhizobium* bacteria rely on the plant in the root rhizosphere for the energy required to break down nitrogen molecules into nitrogen atoms. However, the nitrogenase enzyme is sensitive to oxygen, and to protect it, either high metabolic activity of the *Rhizobium* bacteria or a diffusion barrier is developed at the periphery of the nodule.

During nitrogen fixation, the binding of nitrogen to the enzyme initiates a series of electron transfer reactions. Electrons from ferredoxin are utilized to reduce the iron protein component of the enzyme. The reduced iron protein then binds to ATP, leading to the reduction of the molybdenum-iron protein, another component of the enzyme. This reduction process generates the necessary electrons to break down nitrogen molecules, resulting in the formation of a compound called di-imide (NH_2). This process is repeated twice more, ultimately converting the di-imide into two molecules of ammonia (NH_3). The stepwise representation of this process is as follows:

$$N_2 + 8H + 8e + 16Mg - ATP - 2NH_2 + H2 + 16MgADP + 16P_i$$

The relationship between leguminous plants such as soybean, cowpea, or beans and the bacterium *Rhizobium* is classified as symbiotic. The rhizosphere surrounding the plant roots serves as a protective environment, providing space and acting as an energy source for *Rhizobium* bacteria. Remarkably, these bacteria have the ability to convert gaseous nitrogen from the atmosphere into ammonium (NH_4^+) ions, which are essential minerals for optimal plant growth. In addition to ammonium ions, nitrate (NO_3^-) ions also serve as a principal source of nitrogen for all plants, particularly crop plants, as they contribute to the plant's growth and development. This intricate relationship between a bacterium and a plant exemplifies the remarkable interdependence of two organisms, ultimately benefitting both and serving as a valuable source of energy and food for the human and animal kingdoms alike.

2.6   How is the *Rhizobium* Bacteria Produced En Masse?

Microbial fertilizers are typically formulated as carrier-based preparations that contain highly efficient strains of specific microorganisms or bacteria. When it comes to manufacturing bacterial inoculants, organic matter carrier materials are generally preferred due to their superior efficacy compared to nonorganic matter carrier materials. The production of carrier-based *Rhizobium* microbial fertilizers follows a three-step procedure for mass production. The steps involved in their production are as follows:

1. The microbial strain is first cultured.
2. Next, the carrier material is processed.
3. Lastly, the carrier material is mixed with the broth culture and efficiently packed.

2.7   How is the Microorganism Cultured?

The culture medium used for *Rhizobium* colonies is yeast extract mannitol broth, as described in Table 2.1. This medium is characterized by colorless, translucent, shiny, and small raised colonies. Notably, *Rhizobium* colonies do not absorb the Congo red dye, resulting in a lack of color change. In contrast, colonies of *Agrobacterium* sp. exhibit a red appearance due to the uptake of the Congo red stain.

To prepare the Congo red dye/stain stock solution, 250 milligrams of Congo red stain should be dissolved in 100 ml of water so that the volume comes up to 1 liter. Then the pH of the solution should be adjusted to 6.8. If the medium is intended for isolation purposes, agar should be added accordingly.

Table 2.1 The composition of yeast extract mannitol (YEM) agar medium with Congo red stain

Chemical	Content
Mannitol	10 g
$K_2 HPO_4$	0.5 g
$MgSO_4.7H_2O$	0.2 g
NaCl	0.1 g
Yeast extract	0.5 g
Agar	20 g
Distilled water	1000 ml

2.8   How to Prepare the Inoculum?

The following 10 steps must be utilized to prepare the inoculum:

1. In conical flasks made of Pyrex (250 ml capacity), the YEM broth should be prepared and sterilized.
2. Using an efficient inoculum of *Rhizobium*, the media in the flask must be inoculated, ensuring aseptic/fully sterilized conditions.
3. The incubation should be done at room temperature in an end-to-end rotary shaker (200 rpm) for about a week (5–7 days). After incubation, the population of *Rhizobium* in the broth should be checked, which will serve as a starter culture.
4. Employing the starter culture (at log phase), the contents of the conical Pyrex flask containing 1000 ml of YEM broth (mentioned above) should be inoculated.
5. The YEM broth should be produced in large quantities using a fermentor and then sterilized.
6. After the media has cooled down, the *Rhizobium* inoculum from the 1000 ml conical Pyrex flask mentioned above (at log phase) should be inoculated at a rate of 1–2% inoculum. The culture should be grown by maintaining optimum aeration and continuous stirring of the culture media.
7. The broth is examined again to assess the high population of the inoculated *Rhizobium* organism and to detect any potential contaminants that may have occurred during the growth phase of the microbial organism.
8. Once the cell population reaches 10^9 and cfu ml^{-1} is attained after the incubation period, the cells can be harvested.
9. There should be no fungal or any other bacterial contamination at the 10^{-5} dilution level.
10. After fermentation, the broth should not be stored for more than 24 h.

2.9   Processing the Carrier Material

To ensure the high quality of microbial fertilizers, the selection of an ideal carrier material is crucial. Commonly used carrier materials include vermiculite clay, lignite, peat soil (rich in organic matter/organic carbon), charcoal, farm yard manure (FYM), press mud, and soil-sand mixture. Among them, neutralized peat soil or lignite is considered a superior carrier material for culturing *Rhizobium*. When choosing a carrier material, the following criteria must be met to ensure its suitability:

1. It must be cost-effective and cheap.
2. It must be locally procurable.
3. It must contain a high amount of organic matter/organic carbon.
4. It must not be contaminated with any external/toxic materials, especially toxic chemicals that will adversely affect *Rhizobium* culture growth.
5. It must have more than 50% water-holding capacity (WHC).
6. It should be easy to handle.
7. If the carrier material is either peat or lignite, it must pass through a 21-micron sieve.
8. The pH of the carrier material should be close to neutral, ideally ranging from 6 to 8 (around 7.2, considering that achieving absolute neutrality in solid materials is not feasible).
9. The carrier material should be amenable to autoclaving.

2.10   How to Mix the Carrier Material with the Broth Culture and Subsequent Packing?

It is essential to strictly adhere to the following steps to carry this out:

1. The sterilized carrier material, which has been neutralized and dried, is spread on a clean and dry polythene sheet or plastic tray.
2. The *Rhizobium* culture obtained from the fermentor is added to the sterilized carrier material, and the two are thoroughly mixed either by hand or using a mechanical mixer. It is important to wear clean or sterilized gloves during this process. The culture suspension should be added in a manner that ensures the moisture content in the mixture is approximately 35–40% of the water-holding capacity (WHC), taking into consideration the microbial population and the specific carrier material being used.
3. Curing is achieved by spreading the *Rhizobium* inoculant on a clean floor or polythene sheet and placing it in an open shallow tub or tray. The inoculant should be covered with a polythene sheet and left at ambient (room) temperature for approximately 48 h before packing.

4. The recommended amount for each inoculant packet is 200 g, which is packaged in polythene bags. The bags are then sealed and stored at ambient (room) temperature until they are ready for use.

2.11   Specifications for the Polythene Bag

The following specifications for the polythene bag must be ensured:

1. The bags should be of low-density grade.
2. The thickness of the bag must be about 75 μm.
3. The bag must carry the following markings:

   A. Name of the manufacturer
   B. Name of the product
   C. Microbial strain used in culturing
   D. The crop for which the inoculant is recommended
   E. Manufacturing date
   F. Batch number
   G. Expiry date
   H. Maximum retail price
   I. Address of the manufacturer and storage details

2.12   Storing the Microbial Fertilizer

The following precautions must be strictly observed:

1. The inoculant package must be stored in a cool place, away from direct sunlight or heat, at room temperature or in a refrigerated storage.
2. At the time of packing the inoculant, the carrier material containing the inoculant should be assessed for population count and also 15 days after packing.

What is Azospirillum?
This is another good example of a microbial fertilizer. It is a gram-negative bacterium that exhibits microaerophilic characteristics, motility, and is a non-fermentative nitrogen-fixing bacteria similar to the *Rhizobium* species. Initially, this bacterium was isolated by Beijerinck in 1922 in Brazil from the roots of the plant *Paspalum* and was named *Azotobacter paspali*. However, it was later reclassified as *Spirillum lipoferum*. In 1976, Dobereiner and Day reported the nitrogen-fixing potential of *Spirillum lipoferum* in the roots of certain forage grasses, introducing the term "associative symbiosis" to describe the loose association of nitrogen-fixing ability in other plants by *Spirillum lipoferum*. The taxonomy of this bacterium was further examined by Tarrand et al. in 1978, leading to the designation of the bacterium as *Azospirillum*. *Azospirillum* is a spiral-shaped organism that colonizes both the internal and external regions of plant roots. Due to its microaerophilic nature, isolation of *Azospirillum* requires the use of semisolid malate medium and enrichment techniques.

2.13 Benefits of *Azospirillum*

The *Azospirillum* bacterium offers several significant benefits in terms of enhancing soil fertility that are described below:

1. It enhances plant nutrient uptake and soil water availability by forming a symbiotic relationship with the plant's roots. This results in better vegetative growth and higher crop yields.
2. Practical experience has shown that the use of *Azospirillum* can reduce the requirement for chemical/synthetic fertilizers by up to 25%. This reduction in fertilizer usage can amount to approximately 25–30 kg of nitrogen per hectare.
3. *Azospirillum* is both eco-friendly and cost-effective.
4. *Azospirillum* can be applied in a wide range of crops, such as rice, wheat, maize, millets (such as sorghum, bajra, ragi), sugarcane, forage crops, and fruit crops, making it a versatile microbial fertilizer.

2.14 Details of *Azospirillum*

Taxonomy:

Kingdom: Bacteria
Phylum: Proteobacteria
Class: Alpha-proteobacteria
Order: Rhodospirillales
Family: Rhodospirillaceae
Genus: *Azospirillum*

2.15 Morphology of *Azospirillum*

Azospirillum, among the species of associative nitrogen-fixing bacteria, has been extensively studied and is considered the most widely investigated species. These bacteria are short, gram-negative, and slightly curved in shape. The rediscovery of *Azospirillum* by Johanna Dobereiner in the mid-seventies played a crucial role in the recognition and study of associative nitrogen-fixing bacteria in the root rhizosphere of plants like *Digitaria* and *Zea mays* in Brazil. While *Azospirillum* is commonly found in tropical soils, its occurrence in temperate soils, tundra, and semidesert sites is rare. *Azospirillum* also has the ability to colonize various plants, including crop plants, weeds, and both annual and perennial grasses.

The following are the eight species of *Azospirillum* described so far:

1. *Azospirillum brasilense*
2. *Azospirillum lipoferum*
3. *Azospirillum halopraeferens*
4. *Azospirillum irakense*

5. *Azospirillum largimobile*
6. *A. dobereinerae*
7. *Azospirillum oryzae*
8. *Azospirillum amazonense*

In fact, it is the association of *Azospirillum* with the host legume plant that stimulates the latter through the atmospheric fixation of gaseous nitrogen (N_2) into plant-assimilable ammonium (NH_{4+}) ions. After the bacterial cells die, the ammonium fixed by the bacteria becomes available for the host plant's nutrition. This symbiotic relationship between organisms is a remarkable feat of nature. The NH_{4+} derived from N_2 fixation by *Azospirillum* is released only in small amounts during diazotrophic growth. The bacteria possess a high-affinity NH_{4+} uptake system and efficient assimilation of NH_{4+} by glutamine synthetase, enabling them to utilize most of the fixed nitrogen for their own growth. Additionally, *Azospirillum* has the ability to produce plant growth-promoting substances, particularly auxins, which enhance lateral root formation and root hair enlargement. This ultimately leads to improved nutrient uptake and enhanced water status in the plant. Among the *Azospirillum* strains, the *Azospirillum lipoferum* and *Azospirillum brasilense* strains have been exploited for commercial production (Table 2.2).

Table 2.2 Constituents of the malic acid broth

Constituent	Content
Malic acid	5 g
KOH	4 g
Dipotassium hydrogen orthophosphate	0.5 g
$MgSO_4$	0.2 g
NaCl	0.1 g
$CaCl_2$	0.2 g
NH_4Cl	1.0 g
Fe-EDTA (1.64% w/v aqueous)	4 ml
Trace element solution	2 ml
BTB (0.5% alcoholic solution)	2 ml
Agar	1.75 g
Distilled water	1000 ml
pH	6.8
Trace element solution:	
Sodium molybdate	200 mg
Manganous sulfate	235 mg
Boric acid	280 mg
Copper sulfate	8 mg
Zinc sulfate	24 mg
Distilled water	200 ml

How is the *Azospirillum* Mass Produced?

Steps taken for preparation of the inoculum:

Step 1: The malic acid broth should be prepared in conical Pyrex flasks (250 ml capacity) and sterilized.

Step 2: The contents of the flask are inoculated with an efficient strain of *Azospirillum* under aseptic conditions.

Step 3: The contents in the conical flask should be incubated in an end-to-end rotary shaker at a speed of 200 rpm and at room/ambient temperature for a week. After the incubation period, the *Azospirillum* count should be checked to confirm the development of the starter culture.

Step 4: Using the starter culture (at log phase), the contents of the conical flask containing 1000 ml of malic acid broth should be inoculated.

Step 5: A fermentor can be utilized to prepare the malic acid broth in large quantities and sterilize it. After the media has cooled, the *Azospirillum* inoculum from the 1000 ml conical flask at the logarithmic growth phase is inoculated at a rate of 1–2% inoculum. The *Azospirillum* cells are cultured in the fermentor by maintaining optimal aeration and continuous stirring of the culture media.

Step 6: Once again, the broth is carefully examined to assess the high population of the inoculated organism and to detect any potential contaminants that may have entered during the growth period of the inoculum. This quality check is crucial to ensure the purity and effectiveness of the culture.

Step 7: Once the desired population of bacterial cells, reaching approximately 10^9 cfu/ ml^{-1}, is achieved after the designated incubation period, the cells are ready to be harvested.

Step 8: No contamination, either fungal or bacterial, must be allowed at the 10^{-5} dilution level.

Step 9: After fermentation, the broth should not be stored for more than 24 h.

Processing the Carrier Material

To ensure the production of high-quality microbial fertilizer, the selection of an ideal carrier material is crucial. Some commonly used carrier materials include:

1. Peat (soil)
2. Lignite (clay material)
3. Vermiculite (clay material)
4. Charcoal
5. Sugarcane press mud (obtained from sugarcane crushing)
6. FYM (farm yard manure) and soil mixture

The neutralized peat soil-ignite mixture is more preferable as a carrier for *Azospirillum* production. The ideal carrier material should possess the following attributes:

A. Must be cheap and economical (cost-effective)
B. Must be locally available
C. Must contain high-organic carbon

D. Must not contain any toxic materials
E. Must possess more than 50% water-holding capacity (WHC)
F. Must be easy to process
G. Both peat and lignite, when used as carrier materials, must pass through 212 μ IS sieve.
H. The carrier material should be ideally neutral (pH 7.0).
I. The carrier material should be autoclavable.

2.16   How to Mix the Carrier Material with the Broth and Pack It Up?

For the above, the steps provided below must be followed.

(a) The carrier material, which has been neutralized and sterilized, should be carefully distributed on a clean and dry polythene sheet, ensuring its complete sterility (if not, the polythene sheet should be sterilized as well).
(b) The sterilized carrier material is combined with the bacterial culture (*Azospirillum* culture) obtained from the fermentor and carefully mixed either using a sterilized mechanical mixer or by hand while wearing sterilized gloves.
(c) The culture suspension should be added in a manner that ensures the moisture content of the contents is approximately 35–40% of the water-holding capacity (WHC), taking into consideration the microbial population and the specific type of carrier material employed.
(d) The *Azospirillum* inoculant is subjected to curing by spreading it evenly on a clean floor or polythene sheet, ensuring ample airflow or by placing it in an open tray covered with polythene for approximately 2 days at room temperature. Following curing, the inoculant is carefully packed using high-quality packaging materials that are free from any form of contamination.
(e) The recommended packaging for the *Azospirillum* inoculant is typically in 200-g polythene bags. Once filled, the bags are sealed at room temperature and stored until they are ready for field application.

2.17   What Specifications are Prescribed for the Polythene Bags Used for Packing the Inoculum?

1. The bags must have a thickness of about 70–75 μ.
2. When it comes to polythene bags, there are two types available, high density and low density. For packing the *Azospirillum* inoculum, it is important to use bags made of low-density polythene.
3. The inoculum packet must have the following markings:

(a) Name of the manufacturer
(b) Name of the product
(c) Name of the microbial strain packed
(d) Crop for which the microbial strain is recommended
(e) Method of inoculation
(f) Date of manufacture
(g) Batch number
(h) Expiry date
(i) Maximum retail price
(j) Full address of the manufacturer
(k) Storage instructions

2.18 What is the Best Mode to Store Microbial Fertilizer Packages?

It is crucial to strictly adhere to the following instructions:

1. The inoculant population in the carrier inoculant packet may be assessed at:

   (a) The time of packing
   (b) A fortnight after packing

2. The packet should contain more than 10^9 cfu/g of the inoculant at the time of packing and 10^7 cfu/g on a dry weight basis on the 15th day before the expiry date.
3. The inoculant package should be stored in a cool, well-ventilated area, away from direct heat, sunlight, and fluctuations in temperature. It is important to maintain an ambient temperature in the storage space.

2.19 Details of Azotobacter

Azotobacter is another significant nitrogen-fixing bacteria, second only to *Azospirillum* in importance. Similar to *Azospirillum*, *Azotobacter* is a free-living bacterium capable of atmospheric nitrogen fixation. It forms cysts and is predominantly found in soils with varying pH levels, ranging from neutral to slightly alkaline. This bacterium exhibits a high metabolic rate, enabling the oxygen-sensitive nitrogenase enzyme to function in the presence of limited oxygen (O_2). Additionally, it produces a protective protein that shields the nitrogenase enzyme from oxygen-induced stress. Azotobacter cells are typically large and rod-shaped, occurring singly, in chains, or in clumps, and can exhibit either motility or be stationary. The formation of cysts around the bacterium serves as a protective mechanism against unfavorable external environmental conditions.

The taxonomy of *Azotobacter* is as follows:

(a) Domain: Bacteria
(b) Kingdom: Bacteria
(c) Phylum: Proteobacteria
(d) Class: Gamma-proteobacteria
(e) Order: Pseudomonadales
(f) Family: Pseudomonadaceae/Azobacteraceae
(g) Genus: Azotobacter

2.20 Which are the Most Important Species of *Azotobacter*?

The following are the most important species of *Azotobacter*:

1. *Azotobacter armeniacus*
2. *Azotobacter agilis*
3. *Azotobacter beijerinckii*
4. *Azotobacter chrococcum*
5. *Azotobacter nigricans*
6. *Azotobacter paspali*
7. *Azotobacter salinestris*
8. *Azotobacter tropicalis*
9. *Azotobacter vinelandii*

2.21 What are the Steps Involved in the Mass Production of Azotobacter?

There are, primarily, three steps involved, that is:

1. How to culture the bacterium?
2. What is the composition of the broth employed?
3. How to prepare the inoculum?

2.22 How to Culture the Bacterium?

The broth used for culturing the bacterium en masse is Ashby's mannitol agar. Its composition is described in Table 2.3.

Table 2.3 The composition of Ashby's mannitol agar

Constituent	Content
Mannitol	10 g
Ca CO$_3$	5 g
K$_2$ HPO$_4$	0.5 g
MgSO$_4$ 7H$_2$O	0.2 g
NaCl	0.2 g
Ferric chloride	Trace
MnSO$_4$ 4H$_2$O	Trace
N-free washed agar	15 g
pH	7.0 (neutral)
Distilled water	1000 ml

2.23 How to Prepare the Inoculum?

The following 10 steps have to be taken to prepare the inoculum in a fool-proof manner:

Step 1: The Ashby's mannitol agar broth must be prepared in a Pyrex conical flask of 250 ml capacity and thoroughly sterilized.

Step 2: The broth in the conical flask should be inoculated with a highly efficient strain of *Azotobacter*. It is very important to ensure that it is done under complete aseptic conditions.

Step 3: The incubation should be done in an end-to-end rotary shaker (200 rpm) for a week for the population (of *Azotobacter*), which will serve as a starter culture.

Step 4: Using the starter culture (at log phase), the contents of the conical flask containing 1000 ml of Ashby's mannitol agar broth should be inoculated.

Step 5: The Ashby's mannitol broth is prepared in a fermentor in large quantities and it must be fully sterilized.

Step 6: Subsequent to cooling of the media, the *Azotobacter* inoculum from the 1000 ml conical flask (at log phase) is added to the broth at a 1–2% inoculum concentration. The *Azotobacter* cells are then grown in a fermentor with optimal aeration and continuous stirring of the culture media.

Step 7: Once again, the broth is carefully examined to assess the population density of the inoculated organism and to detect any potential contaminants that may have been introduced during the growth phase of the inoculum.

Step 8: Once the desired population of the inoculum of 10^9cfu ml^{-1} is attained after the incubation period, the bacterial cells can be harvested.

Step 9: No fungal or bacterial contamination should be permitted at the 10^{-5} dilution level to ensure the purity of the sample.

Step 10: The broth, after proper fermentation, should not be stored for more than 24 h to prevent significant degradation of the cell mass caused by environmental or other contaminations.

2.24   How to Process the Carrier Material?

The selection of an efficient and suitable carrier material is crucial to ensure the production of high-quality microbial fertilizers. The following materials can be considered for use as a carrier material:

(a) Peat (soil)
(b) Lignite (clay material)
(c) Vermiculite (clay material)
(d) Sugarcane press mud (a by-product of sugarcane crushing)
(e) Charcoal
(f) FYM (farm yard manure) and soil mixture

Ideally, the peat (soil)-lignite mixture is a better carrier material for *Azotobacter* culture than the others due to its favorable characteristics.

The ideal carrier material should conform to the following specifications:

1. Must be locally available
2. Must be inexpensive and cost-effective
3. Must contain large quantities of organic carbon (organic matter)
4. Must not contain any toxic chemicals and/or other contaminants
5. Must have more than 50% water-holding capacity (WHC)
6. Must be amenable for easy processing
7. When the carrier material is either peat or lignite, each must pass through a 212 micron IS sieve.
8. The carrier material must be neutral in reaction (pH 7.0).
9. The durability of the carrier material should also be ensured.

How to Ensure Proper Mixing of the Carrier Material with the Culture Broth and Subsequent Packing?

It is expedient to strictly adhere to the following points:

1. The sterilized and neutralized carrier material should be spread on a thoroughly sterilized, clean plastic tray or alternatively, a sterilized metallic tray can also be used.
2. The sterilized carrier material should be mixed thoroughly with the *Azotobacter* culture obtained from the fermentor. This can be done either manually, using a thoroughly sterilized glove, or by using a mechanical mixer to ensure proper mixing.
3. The culture suspension should be added in such a way that the moisture content is about 35–40% of water-holding capacity (WHC). The exact amount needed may vary depending on the microbial population and the carrier material used.
4. The *Azotobacter* culture is evenly spread on a clean floor or a clean polythene sheet using gloved hands. It is then placed in sterilized water tubs that have been thoroughly dried. The culture is covered with a clean, sterilized polythene sheet and left in a well-ventilated and clean room at ambient temperature for a period

of 2–3 days. This curing process ensures the proper development and stabilization of the *Azotobacter* culture before it is packed and dispatched for use or further storage.
5. The recommended inoculant packet size is 200 g and it should be packed in polythene bags, which are then sealed at room temperature.

Microbial Fertilizers: The Example of Phosphatic Fertilizers
These microbial fertilizers are capable of solubilizing or mobilizing phosphate ions, including the most difficult ones found in the soil (Nair 1996). This process makes phosphate readily available to the plants for uptake and utilization.

2.25 The Illustrious Example of Phosphate-Solubilizing Microorganisms (PSM): How do These Microorganisms Function?

The challenges associated with phosphorus bioavailability, which is considered the most difficult major nutrient among nitrogen, phosphorus, and potassium for optimal plant nutrition, have been extensively discussed by Nair (1996). When it comes to phosphorus uptake by plant roots, it is absorbed in the form of $H_2PO_4^-$ or HPO_4^{2-}. Phosphorus plays a crucial role in storage and energy transfer within the plant system, as well as in various physiological functions such as photosynthesis, carbon metabolism, and membrane formation.

The efficiency of phosphate fertilizer utilization is typically low, ranging from 10% to 25%, and the concentration of plant/soil available phosphorus at any given time is relatively low. The fixation of applied phosphorus in both acidic and alkaline soils is a major challenge in plant nutrition. However, certain specific microorganisms have the ability to solubilize "fixed insoluble phosphate" and convert it into a bioavailable form, making it accessible to plant roots.

2.26 Making Fixed Phosphate Bioavailable Through Which Route

Microorganisms that possess the ability to solubilize "fixed" phosphate are known as "phosphate-solubilizing microorganisms" (PSMs). This solubilization process is facilitated by the action of two enzymes, (1) phytase and (2) phosphatase, along with the secretion of organic acids like gluconic, lactic, and citric acids. Some of the primary PSMs include the following:

1. *Pseudomonas*
2. *Mycobacterium*
3. *Micrococcus*
4. *Flavobacterium (bacteria)*
5. *Penicillium*
6. *Sclerotium*

Table 2.4 The important phosphate-solubilizing bacteria (PSB) and their metabolic forms

Name of the PSB	Acidic metabolite form
Aspergillus niger, Penicillium sp.	Lactic acid
E. freundii	Lactic acid
Penicillium rugulosum	Citric and gluconic acids
Enterobacter intermedium	2-keto gluconic acid
Aspergillus flavus, Aspergillus niger, Penicillium canescens	Oxalic, gluconic, citric, succinic acids
Aspergillus niger	*Oxalic, gluconic*
P. fluorescens	Citric, malic, tartaric, gluconic
Arthrobacter sp.	Citric, malic, tartaric, gluconic
Enterobacter sp.	Citric, malic, tartaric, gluconic, fumaric
P. trivialis	Lactic, formic
Pseudomonas putida M5TSA, *Enterobacter*	Malic, gluconic
sakazakii, M2PFe and *Bacillus megaterium* MIPCa	Malic, gluconic
Enterobacter sps Fs 11	Malic, gluconic
Aspergillus niger FS 1, Penicillium canescensFS23	Citric, gluconic, oxalic
Eupenicillium ludwigii FS27	Citric, gluconic, oxalic
Penicillium islandicum FS30	Citric, gluconic, oxalic
Pseudomonas nitroreducens	Indole acetic acid

7. *Aspergillus* (fungus)

During the solubilization process, a portion of the solubilized phosphorus is utilized by the microorganism itself. In many ways, it resembles the remarkable symbiotic relationship found in atmospheric nitrogen fixation, as discussed earlier in this book.

Phosphorus undergoes both mineralization (solubilization) and immobilization within the soil matrix. These processes are influenced by the quantity of phosphorus present in decomposing plant residues and the nutrients necessary for the associated microbial population.

Table 2.4 provides a comprehensive overview of the important PSMs and their respective metabolic forms.

2.27 Producing PSM En Masse

To produce PSM on a large scale, the following steps should be carried out:

Step 1: How to culture the PSM?

The Pikovskaya's broth is commonly used as a medium to culture the PSM. The details of this broth are given in the following table (Table 2.5).

Table 2.5 The composition of the Pikovskaya's culture broth

Constituent	Content
Glucose	10 g
Calcium phosphate, $Ca_3 PO_4$	2.5 g
Ammonium sulfate, $NH_4 SO_4$	0.5 g
Potassium chloride, KCl	0.2 g
Magnesium sulfate, $Mg SO_4 7H_2O$	0.1 g
Manganese sulfate, $MnSO_4$	Trace
Iron sulfate, $FeSO_4$	Trace
Yeast extract	0.5 g
Distilled water	1000 ml

2.28 How to Prepare the Inoculum?

The following 10 steps should be utilized to prepare the inoculum:

Step 1: In conical Pyrex flasks (250 ml capacity), the Pikovskaya's broth should be prepared and sterilized.

Step 2: The above media should be inoculated with an efficient PSM strain under complete aseptic conditions.

Step 3: The contents in the conical flask must be incubated in an end-to-end rotary shaker (200 rpm) for about a week and the population should be checked for the presence of the PSM. This will serve as the starter culture.

Step 4: Using the starter culture (at log phase), the contents of the Pyrex conical flask containing 1000 ml of Pikovskaya's broth must be inoculated.

Step 5: The Pikovskaya's broth should be prepared in large quantities in a fermentor and sterilized.

Step 6: After the media has cooled down, the PSM inoculum from the 1000 ml conical flask (log phase) is added at a rate of 1–2% inoculum. The PSM cells are cultivated in a fermentor with continuous stirring and optimal aeration of the culture media.

Step 7: Additionally, the broth is carefully examined to ensure a high population of the PSM inoculum and to detect any potential contaminants that may have entered during the PSM growth process.

Step 8: Once the cell population reaches 10^9 ml^{-1} after the designated incubation period, the cells can be harvested.

Step 9: It is crucial to ensure that there is no fungal or bacterial growth observed at a dilution level of 10^{-5}.

Step 10: Following fermentation, the broth should not be stored for more than 24 h.

2.29   How to Process the Carrier Material and Which are the Preferred Carrier Materials?

(a) Peat (soil)
(b) Clay (lignite, vermiculite)
(c) Charcoal
(d) Sugarcane press mud (a by-product of sugarcane crushing)
(e) FYM (farm yard manure)
(f) Soil-sand mixture

Compared to the other materials, the neutralized peat (soil)-lignite mixture is a better carrier material for PSM production.

To properly process the carrier material, the following nine characteristics have to be present:

1. The carrier material must be locally available.
2. The carrier material must be cheap and cost-effective.
3. The carrier material must contain high amounts of organic carbon (organic matter).
4. The carrier material should not contain any toxic materials and/or contaminants, which will adversely affect the growth of the inoculum.
5. It must have more than 90% of water-holding capacity (WHC).
6. The carrier material must be easy to process.
7. When peat/lignite is used as the carrier material, it must pass through 212 micron IS sieve.
8. The pH of the carrier material must be neutral (7.0).
9. The carrier material used must be autoclavable.

2.30   How to Mix the Carrier Material with the Culture Broth and Pack It?

The following five steps should be performed:

Step a: The neutralized and sterilized carrier material should be spread evenly on a very clean, sterilized, and dry plastic tray.

Step b: The culture (PSM) obtained from the fermentor must be added to the sterilized carrier material and thoroughly mixed manually, wearing sterilized gloves or in an end-to-end shaker (at rpm 200).

Step c: The culture suspension should be added in such a manner that the moisture is maintained at 35–40% water-holding capacity (WHC), based on the microbial population and the type of carrier material used.

Step d: The PSM inoculant must be spread on a very clean floor/polythene sheet by keeping in an open shallow bath tub/tray with polythene covering for about 2 days at ambient temperature before packing so that the material can be properly cured.

Step e: The typical inoculant package weighs 200 g and is enclosed in polythene bags, which are sealed at room temperature.

2.31   What are the Specifications Used for the Polythene Bags?

The following specifications must be strictly adhered to when it comes to polythene bags:

1. The polythene bag must be of low-density grade.
2. Ideally, they must be 50–75 microns thick.
3. The packet must carry the following details:

   (a) Manufacturer's name
   (b) Product name
   (c) Microbial strain used
   (d) Recommended crop
   (e) Inoculation method
   (f) Manufacture date
   (g) Batch number
   (h) Expiry date
   (i) Maximum retail price (MRP)
   (j) Full address of the manufacturer
   (k) Proper instructions for storage

2.32   How to Store the Microbial Fertilizer Properly?

The following two instructions must be strictly adhered to:

1. To store the inoculant, a cool storage area that is away from direct sunlight and heat should be chosen. This storage space must be well ventilated and, preferably, a cold storage facility.
2. The population of the inoculant in the carrier packet can be determined either at the time of packing or 2 weeks prior to the expiry date. It is recommended to have a population of more than 10^9 cfu/g of inoculant at the time of packing and 10^7 cfu/g based on the dry weight 15 days before the expiry date.

2.33   Phosphate-Mobilizing Fungus (PMF)

In addition to bacteria, mycorrhizal fungi also form a symbiotic association with plant roots. In this association, the plant provides carbon sources to the fungi, and in return, the fungi provide important mineral nutrients to the plant, promoting vegetative growth and reproduction. This symbiotic association is similar to the association between *Rhizobium/Azospirillum/Azotobacter* and legume plant roots, where the plant contributes by fixing atmospheric elemental nitrogen (N_2). In this discussion, we will focus on the solubilization of "fixed" soil phosphorus into plant-available forms. Mycorrhizal fungi have the ability to absorb, accumulate, and transport large quantities of phosphate within the plant system, primarily within their hyphae, and release it to the plant root tissues. Mycorrhizal fungi are endophytic, biotrophic, and mutalistic symbionts found in many cultivated and natural ecosystems. There are three major groups of mycorrhizal fungi:

1. Ectomycorrhiza
2. Endomycorrhiza
3. Ectendomycorrhiza

Each of these groups has specific functions in crop production.

Ectendomycorrhiza and Endomycorrhiza are particularly important in agricultural forestry. In the past, agriculture and forestry were separate domains. However, the concept of "agricultural forestry" emerged in the mid-twentieth century, combining annual crop plants with perennial forest trees. This development had a significant impact, exemplified by the establishment of the International Centre for Research in Agroforestry (ICRAF) in Nairobi, Kenya, under the Consultative Group for International Agricultural Research (CGIAR), primarily funded by the United States. Endomycorrhiza, commonly known as vesicular arbuscular mycorrhiza (VAM), enhances plant growth and yield by improving nutrient supply to the host plant. VAM fungi are widely used as inoculants in phosphorus-deficient soils where other soil amendment practices may fail. However, the "Nutrient Buffer Power Concept," developed by the author (Nair 1996), plays a crucial role in resolving soil-related phosphorus issues. VAM fungi can supplement the concept in such cases. Mycorrhizal plants have the ability to absorb and accumulate several times more phosphate from phosphorus-deficient soils compared to non-mycorrhizal plants. Endomycorrhiza, specifically VAM fungi belonging to the phylum Glomeromycota, form symbiotic relationships with terrestrial plant roots. These fungi not only improve phosphate nutrition in crop plants but also enhance the overall growth of the host plant, making it more resilient to drought and disease. Therefore, VAM fungi hold great potential for sustainable agriculture.

2.34   Benefits of Mycorrhiza

The mycorrhizal association provides several important benefits, including:

1. Enhanced phosphorus absorption: Mycorrhiza plays a crucial role in improving the absorption of phosphorus by plant roots.
2. Increased plant growth and reproduction: The symbiotic relationship with mycorrhizal fungi leads to better plant growth and reproduction, resulting in higher yields.
3. Enhanced resistance to pests, diseases, and drought: Mycorrhizal association enhances the resistance capacity of host plants to various pests and diseases, particularly root and collar rot diseases. It also imparts resistance to drought stress.

Taxonomy of Vesicular Arbuscular Mycorrhiza (VAM) Fungi

VAM fungi, also known as arbuscular mycorrhizal fungi, have specific characteristics and taxonomy. The spores of VAM fungi range in size from 50 to 500 μ and are larger than typical fungi. These spores are usually formed within the soil matrix. Unlike other fungi, VAM fungi lack septa in their hyphae since they do not have a sexual phase of growth.

Under favorable conditions, VAM spores germinate and extend their hyphae into plant roots. The hyphae penetrate the cortex layer of the roots, forming vesicles or arbuscules, or sometimes both, which are distinctive features of VAM fungi. It should be noted that VAM fungi of the Gigasporaceae family do not form internal vesicles in roots.

The colonization of VAM fungi on plant roots is essential for their proliferation, and they are considered "obligate symbionts." The interaction between VAM fungi and plants is based on mutualism, involving nutrient exchange, similar to the symbiotic association between Rhizobium and leguminous plants like peas (Pisum sativum), as discussed earlier in this book. Recent molecular phylogenetic studies have revealed that the traditional classification of Zygomycota is polyphyletic. A new phylum called "Glomeromycota" was proposed for VAM fungi, although they were previously considered members of the Zygomycota family. While molecular identification methods are gaining popularity, conventional morphological observations remain valid and should not be overlooked for accurate identification of these fungi.

2.35   The Presence of VAM Fungi on Plant Roots

To identify VAM fungi structures in plant root samples, staining techniques can be used (Philips and Hayman 1970). Freshly collected root samples should be thoroughly washed with distilled or clean water to remove any soil particles. If tightly adhering soil particles are present, an ultrasonic technique can be used to disperse them. The stained roots can then be examined under a dissecting microscope or a compound microscope with transmitted illumination. Fungal structures such as vesicles and arbuscules can be easily recognized.

How to Isolate the VAM Spores?

VAM spores can be isolated and collected from root samples using the wet-sieving and decanting technique (Gerdmann and Nicolson 1963). Spores are lighter in gravity compared to soil particles. By repeatedly decanting the soil suspension and wet sieving it through a fine mesh sieve, the spores can be concentrated from the soil sample. Under a dissecting microscope, the spores can be identified as globular or sub-globular, with a diameter of 50–500 μ.

2.36   How to Identify the Morphological Characteristics of the VAM Spores?

Morphological characteristics play a crucial role in the identification of VAM spores. However, spores collected from the field may decay, making identification tentative and limited to the genus level. It is recommended to observe a minimum of 30–50 spores of the same morphological type under a stereomicroscope and a compound microscope. The morphological characters such as shape, size, color, and spore attachment should be carefully observed under the microscope. Based on these characteristics, the spores can be classified into different types. Detailed observations are necessary for each spore type.

2.37   Observing the Spore Under a Compound Microscope

To observe multiple spores of each type, mount them on a glass slide with polyvinyl lactoglycerol (PVLG) medium. Here are the steps to prepare the PVLG medium:

1. Dissolve 1.66 g of polyvinyl alcohol (polymerization 100–1500) in 10 ml of double-distilled water.
2. Allow the polyvinyl alcohol to completely dissolve, which may take around 6 h at 80 degrees Celsius (°C).
3. Mix the dissolved polyvinyl alcohol with 1 ml of lactic acid and 1 ml of glycerol.

For certain genera of VAM fungi, such as Archaeospora, molecular data may be required along with morphological characteristics for identification. To identify species, the characteristics should be compared with the descriptions available in the original references. Detailed descriptions of many species can be found on the INVAM website. For measuring spore size, it is recommended to examine a minimum of 40–50 spores.

The characteristics provided in Table 2.6 will help identification of the spore at the level of the genus.

Table 2.6 Characteristics required for spore identification at genus level

Characteristics	Remarks
Shape	Globular, spherical, irregular
Size	Globular: diameter (minimum–average–maximum Irregular shape: length x width (minimum–average–maximum)
Color	Compare with standard color chart
Hyphal attachment	Soporiferous saccule, bulbous suspensor
Soporiferous succule	Observed in Acaulospora, Entrophospora, Archaeospora
Bulbous suspensor	Observed in Gigaspora, Scutellospora
Auxiliary cell	Observed in Gigaspora, Scutellospora
Sporocarp	Present, absent
Germination shield	Present, Scutellospora, others, absent
Surface ornamentation	Smooth, rough, reticulate
Vesicle	Presence, absence of mycorrhizal roots

2.38   How to Produce VAM Fungi En Masse?

Spore culture method:

The following procedure can be adopted if the following conditions are met:

Condition 1: High VAM colonization in a plant community with limited or no spore production, especially in arid and hydric conditions

Condition 2: High microbial activity in the soil, particularly in tropical environments with high humidity, temperature, and soil moisture

Condition 3: Despite the presence of high organic matter content, AMF spores may not readily transform.

Here is the procedure to be followed:

Step 1: Collect the rhizosphere soil along with the shoots of the trap plant and cut the plant at the crown. Finely chop the roots and thoroughly mix them with the soil.

Step 2: Mix the chopped roots along with the soil in a 1:1 (v/v) ratio with autoclaved coarse sand in a plastic bag.

Step 3: Transfer the mixture to a plastic pot with a diameter of 15 cm.

Step 4: Sow seeds of the trap plant, such as maize, sorghum, or onion, in the plastic pot.

Step 5: Transfer the pots to a greenhouse and maintain them for around 45 days, monitoring the sporulation of AMF regularly. After approximately 60 days, AMF sporulation may reach its peak. It is crucial to ensure that there is no contamination in the pots.

Step 6: Apply small quantities of chemical fertilizers to encourage AMF proliferation.

Step 7: Once the spore count and percent root colonization have increased, leave the trap culture pots to dry under shade for 2 weeks.

Step 8: Harvest the spores using the sieving and decanting technique, as described by Gerdmann and Nicolson (1963), and determine the spore count.

Step 9: Use the pure AMF culture obtained from the above eight steps to colonize the roots of seedlings of the chosen host crop.

2.39　How to Produce AMF En Masse on the Farm Front?

The following steps should be performed for the mass production of AMF on the farm:

Step 1: Dig a trench of size 1 square meter × 30 cm (length, breadth, and depth) on the farm soil at a convenient location.

Step 2: Line the trench with a black polythene sheet and fill it with a mixture of vermiculite and coir pith up to a height of 20 cm.

Step 3: Apply AMF inoculum at a rate of 2 kg/square meter and spread it uniformly to a depth of approximately 5 cm.

Step 4: Sow seeds of host/trap crops, such as maize, onion, or sorghum. Ensure that the seeds are thoroughly sterilized.

Step 5: Apply synthetic fertilizers (urea, superphosphate, and muriate of potash) in a ratio of 2:2:1 to the mixture in the trench.

Step 6: Allow the host/trap crops to grow for about 2 months. Mix the plant roots along with the vermiculite/coir pith to obtain bulk AMF inoculum.

Step 7: Store the AMF inoculum in sterilized polybags in a well-ventilated storage room so that it remains viable for up to 180 days.

2.40　Potash-Solubilizing Bacterium (KSB)

Among the major plant nutrients required for all crops, nitrogen (N), phosphorus (P), and potassium (K), the last is not the least important. This is because in many situations, potassium plays a crucial role in plant nutrition. Though in quite a few soils, potassium can be found abundant, only about 2% of it is bioavailable to the crop plant. The dynamics of potassium bioavailability is thoroughly discussed by Nair (2019a, b) in the book on intelligent soil management.

Potassium-solubilizing microbes/bacteria improve the growth of the plant and its yield. These microbes solubilize K from the inorganic and insoluble pools to soluble ones to enable plant root to absorb it. Higher K-solubilizing bacteria are generally found in the rhizosphere than in the non-rhizosphere soil. Bacterial isolates having K-solubilizing potential can be screened by using modified Aleksandrov's medium, which is mainly based on halo-zone formation surrounding the bacterial colonies. These microbial colonies are as follows:

1. *Frateuria aurantia*
2. *Bacillus mucilaginosus*
3. *Bacillus edaphicus*
4. *Bacillus circulans*

These can be considered as efficient K solubilizers.

2.41   How Does the Mechanism of K Solubilization Operate?

K solubilization and mobilization are facilitated by the production of various types of organic acids by potassium solubilizing bacteria (KSB). These organic acids play a crucial role in converting insoluble potassium from minerals like mica, muscovite, biotite, and feldspar into a soluble form in the soil solution, thereby increasing its bioavailability. The specific types of organic acids produced by different bacteria are provided in Table 2.7.

In addition to the above, the following bacteria can also function as KSB:

1. *Bacillus edaphicus*
2. *Bacillus circulans*
3. *Paenibacillus* sp.
4. *Acidithiobacillus ferrooxidans*
5. *Pseudomonas*
6. *Burkholderia*

2.42   The Bacteria That Solubilize Zinc and Silica

Among micronutrients, zinc holds significant importance. Due to the widespread adoption of chemical-intensive farming practices, popularly known as the "green revolution," many soils in South Asia, particularly in India and Pakistan, have

Table 2.7 Details of KSB

Bacteria	Organic acid produced
Bacillus mucilaginosus	Oxalic and citric
Pseudomonas sp.	Tartaric and citric
Pseudomonas aeruginosa	Acetic, citric, oxalic
Paenibacillus mucilaginosus	Tartaric, oxalic, citric
Enterobacter asburiae and *Bacillus metallica*	Lactic, gluconic
Bacillus megaterium, *Pseudomonas* sp. and *Bacillus subtilis*	lactic, malic, and oxalic
Bacillus megaterium, *Enterobacter freundii*	Citric, gluconic
Arthrobacter sp., *Bacillus* sp., and *Bacillus firmus*	Lactic, citric

become severely zinc deficient over the past few decades. The bioavailability of zinc and its management techniques have been extensively discussed by Nair (1996, 2019a, b), who introduced the innovative concept of "The Nutrient Buffer Power Concept" for enhancing zinc bioavailability globally. Zinc-solubilizing microbes play a crucial role in solubilizing soil zinc through acidification of the soil. These microbes release organic acids that break the bond between zinc and the soil matrix, thus increasing its availability. Acidification of the soil matrix also leads to a reduction in soil pH, which aids in zinc solubilization. Anions can also chelate zinc and enhance its bioavailability. Additional mechanisms involved in zinc solubilization include the production of siderophores, proton and oxidoreductive systems on cell membranes, and chelated ligands. The following bacteria possess the ability to solubilize soil zinc:

1. *Pseudomonas aeruginosa*
2. *Gluconacetobacter diazotrophicus*
3. *Bacillus* sp.
4. *Pseudomonas striata*
5. *Pseudomonas fluorescence*
6. *Burkholderia cenocepacia*
7. *Serratia liquefaciens*
8. *Serratia marcescens*
9. *Bacillus thuringiensis*

It is worth noting that the bacterium *Bacillus thuringiensis* (listed as item no. 9) has gained notoriety due to its involvement in the development of genetically modified (GM) cotton, commercially known as "Bollgard," by Monsanto through RNA gene transfer techniques. The introduction of Bollgard cotton in India faced significant challenges as the targeted pink bollworm pest developed resistance, leading to crop failure and subsequent hardships for cotton farmers. The author had warned the Indian agricultural community about the potential failures of Bollgard cotton due to the pest's ability to mutate and the emergence of new pests. Unfortunately, the dire predictions came true, resulting in negative consequences for Indian agriculture and the farming community. The Bollgard episode, coupled with the negative impacts of the green revolution, caused immense distress, including the tragic occurrence of mass farmer suicides in regions like Vidarbha in Maharashtra. The high financial burden of investing in Bollgard seeds and chemical fertilizers, coupled with the increased input costs for irrigation and chemicals, pushed many cotton farmers into bankruptcy. The Indian agricultural sector witnessed an unprecedented number of farmer suicides as a result of these challenges. This unfortunate episode remains a dark chapter in Indian agriculture, highlighting the consequences of unsustainable farming practices and the influence of vested interests in the industry.

2.43   How is Silica Solubilized?

The following bacteria have the capacity to solubilize silicate rocks:

1. *Bacillus mucilaginosus*
2. *Bacillus circulans*
3. *Bacillus edaphicus*
4. *Burkholderia*
5. *Arthrobacter ferrooxidans*
6. *Arthrobacter* sp.
7. *Enterobacter hormaechei*
8. *Paenibacillus mucilaginosus*
9. *Paenibacillus frequentans*
10. *Cladosporium*
11. *Aminobacter*
12. *Sphingomonas*
13. *Paenibacillus glucanolyticus*

2.44   What are the Plant Growth-Promoting Bacteria (PGPR)?

PGPR refers to bacteria that have a beneficial effect on plant growth. These bacteria belong to the following genera:

1. *Actinoplanes*
2. *Agrobacterium*
3. *Alcaligenes*
4. *Amorphosporangium*
5. *Arthrobacter*
6. *Azotobacter*
7. *Bacillus*
8. *Cellulomonas*
9. *Enterobacter*
10. *Erwinia*
11. *Flavobacterium*
12. *Pseudomonas*
13. *Rhizobium*
14. *Bradyrhizobium*
15. *Streptomyces*
16. *Xanthomonas*

All of the abovementioned PGPR bacteria promote optimal plant growth, leading to higher yields of the respective plants/crops. This significantly benefits the farmers and entrepreneurs involved in their cultivation.

References

Gerdmann JW, Nicolson TH (1963) Spores and mycorrhizal endogone species extracted from soil by wet-sieving and decanting. Trans Br Mycol Soc 46:235–244

Nair KPP (1996) The buffering power of plant nutrients and effects on availability. Adv Agron 57:237–287

Nair KPP (2019a) Intelligent soil management for sustainable agriculture – the nutrient buffer power concept. Springer Nature Switzerland AG

Nair KPP (2019b) Combating global warming the role of crop wild relatives for food security. Springer Nature Switzerland AG

Phillips JM, Hayman DS (1970) Improved procedures for clearing roots and staining parasitic and vesicular-arbuscular mycorrhizal fungi for rapid assessment of infection. Trans Br Mycol Soc 55:158–161

Chapter 3
How do Microbial Fertilizers Function and How is the Efficiency of Microbial Fertilizers Quantified?

Abstract The chapter will elaborately discuss the functioning of various microbial fertilizers and the manner in which their efficiency can be quantified. It will also discuss the merits of multi-strain inoculants and a consortium of microbial fertilizers.

Keywords Microbial fertilizer efficiency · Multi-strain inoculants · Consortium of microbial fertilizers

The primary mechanism by which microbial fertilizers function in the soil is by modulating the microbial population in the root rhizosphere of the specific plant/crop. The microbial fertilizers facilitate plant growth by promoting nutrient absorption and utilization or by secreting growth-promoting hormones. Chapter 2 of this book explains the processes of phosphorus, potassium, zinc, and silica solubilization. Additionally, microbial fertilizers can have a beneficial effect on plant growth by acting as biocontrol agents against various plant pathogens. The mode of action of these fertilizers can be direct or indirect, resulting in the inhibition of plant/crop growth.

A. **Which are the Direct Mode of Action?**

1. The phenomenon of N fixation.

    All crop plants require an ample supply of nitrogen for optimal growth and yield. Nitrogen, the most crucial macronutrient, is abundantly available in the atmosphere, constituting around 78–80% of it. However, this atmospheric nitrogen is mostly inaccessible to cereal crops such as rice, wheat, maize, etc. Only leguminous plants, belonging to the bean group, possess the ability to fix atmospheric nitrogen through symbiotic nitrogen fixation, as explained in Chap. 2. Biological nitrogen fixation is considered a cost-effective and environmentally friendly alternative to the use of synthetic fertilizers.

© The Author(s), under exclusive license to Springer Nature
Switzerland AG 2023
K. P. Nair, *Extractive Farming or Bio Farming?*, SpringerBriefs in
Environmental Science, https://doi.org/10.1007/978-3-031-34695-8_3

2. How are the nitrogen-fixing bacteria classified?

Nitrogen-fixing bacteria are classified into the following two groups:

(a) Symbiotic nitrogen-fixing bacteria: These bacteria establish a symbiotic association with leguminous plants, such as peas (*Pisum sativum*), with *Rhizobium* being the common symbiont. Additionally, nonleguminous plants like certain trees can form a symbiotic relationship with bacteria, for example, Frankia.

(b) Nonsymbiotic nitrogen-fixing bacteria: This group includes free-living, associative, and endophytic bacteria. Examples of nonsymbiotic nitrogen-fixing bacteria include *Cyanobacteria* (*Anabaena, Nostoc*), *Azospirillum, Azotobacter, Gluconacetobacter diazotrophicus*, and *Azocarus*. However, these bacteria fix only a small quantity of atmospheric nitrogen.

Symbiotic nitrogen-fixing bacteria infect and establish a symbiotic relationship with the roots of leguminous plants. This symbiotic interaction involves a complex interplay between the leguminous host plant and the symbiont bacteria (e.g., Rhizobium). The result is the formation of root nodules where the bacteria (rhizobia) reside as intracellular symbionts. In the case of nonleguminous plants like certain trees (e.g., Frankia), the bacteria, known as diazotrophs, fix atmospheric nitrogen without a symbiotic association. The process of nitrogen fixation is facilitated by a complex enzyme called nitrogenase. Molybdenum nitrogenase is the predominant form enabling biological nitrogen fixation, serving as a vital nutrient for all diazotrophs.

3.1 Enhancing Phosphate Solubilization Through Phosphate-Solubilizing Bacteria (PSB)

Plant roots can absorb phosphorus only in the form of $H_2PO_4^-$ or HPO_4^{2-}. Phosphate-solubilizing bacteria (PSB) play a crucial role in solubilizing soil phosphates. Several PSB species are particularly significant in this process, including the following examples:

1. *Azotobacter*
2. *Bacillus*
3. *Beijerinckia*
4. *Burkholderia*
5. *Enterobacter*
6. *Erwinia*
7. *Flavobacterium*
8. *Microbacterium*

  9. *Pseudomonas*
 10. *Rhizobium*
 11. *Serratia*

Organic phosphorus is mineralized by bacteria through the synthesis of various phosphatase enzymes, which catalyze the hydrolysis of phosphate esters. In certain strains of phosphate-solubilizing bacteria (PSB), both phosphate solubilization and mineralization processes can coexist. Apart from supplying bioavailable phosphorus to plant roots, PSB can also enhance plant growth by stimulating biological nitrogen fixation (BNF) and synthesizing essential plant growth-promoting substances that enhance the bioavailability of other trace elements.

3.2   How are Siderophores Produced?

Microorganisms, including plants, animals, and humans, require the essential micronutrient iron (Fe) for the proper functioning of vital systems. Under aerobic conditions, iron exists in its trivalent form due to oxidative reactions in the soil. However, plants can only absorb iron in its divalent form, which occurs through reduction in waterlogged conditions. The trivalent form of iron forms insoluble hydroxides and oxyhydroxides, making it inaccessible to both plants and microorganisms.

Bacteria have the ability to absorb iron through the secretion of low-molecular-weight iron chelators called siderophores. Siderophores can be categorized as extracellular or intracellular and are generally water soluble. In both gram-positive and gram-negative rhizobacteria, iron (Fe^{3+}) on the bacterial membrane is converted to Fe^{2+}, which enters the plant cell from the siderophore via a gating mechanism linking the inner and outer membranes. Siderophores also act as solubilizing agents for mineral nutrients in the soil, such as iron, as well as for other organic compounds under iron-stressed conditions. Furthermore, siderophores can form stable complexes with other heavy metals, including copper (Cu), manganese (Mn), zinc (Zn), aluminum (Al), cadmium (Cd), and gallium (Ga), thus increasing the soluble metal concentration. This capability of bacterial siderophores to overcome stresses caused by high soil contents of heavy metals is an important factor, especially in the context of intensive farming practices associated with the "green revolution."

Crop plants uptake iron released by bacterial siderophores through a process called chelation. The direct uptake of siderophore-iron complexes occurs via a ligand exchange mechanism. Numerous studies have reported the promotion of plant growth through siderophore-mediated iron uptake by plant roots following bacterial inoculation, as documented in scientific literature.

3.3 Phytohormones: A Very Vital Product of Microbial Activity

Auxin, specifically indole-3-acetic acid (IAA), is a vital product released by microbial activity, particularly by rhizobacteria. It plays a significant role in plant development by influencing changes in the endogenous IAA levels through the acquisition of IAA secreted by soil bacteria. IAA is involved in various aspects of plant growth and defense responses. It enhances the surface area of plant roots, enabling them to capture more soil nutrients. Additionally, IAA softens plant cell walls, facilitating increased root exudation to support the growth of rhizosphere bacteria.

The release of IAA by bacteria has been identified as an effector molecule in plant–microbe interactions, influencing pathogenesis and phytostimulation. Tryptophan, a precursor for IAA synthesis, is a crucial molecule that modulates the level of IAA biosynthesis. Tryptophan stimulates IAA production, while the synthesis of anthranilate, a precursor for tryptophan, reduces IAA synthesis.

3.4 ACC Deaminase (1-Aminocyclopropane-1-Carboxylate Deaminase)

Some microbial fertilizers produce an enzyme called 1-aminocyclopropane-1-carboxylate (ACC) deaminase. This enzyme plays a crucial role in enhancing plant growth, development, and yield by regulating the concentration of ethylene, which can hinder various metabolic activities. Several bacterial strains exhibit this beneficial phenomenon, including the following examples:

1. *Acinetobacter*
2. *Achromobacter*
3. *Agrobacterium*
4. *Alcaligenes*
5. *Azospirillum*
6. *Bacillus*
7. *Burkholderia*
8. *Enterobacter*
9. *Pseudomonas*
10. *Ralstonia*
11. *Serratia*
12. *Rhizobium*

The abovementioned rhizobacteria possess the ability to take up the ethylene precursor ACC and convert it into 2-oxobutanoate and ammonia. ACC deaminase-producing bacteria have been found to alleviate various forms of plant stress, including stress caused by phytopathogenic microorganisms (viruses, bacteria, fungi, etc.), polyaromatic hydrocarbons, heavy metals, radiation, wounding, insect predation, high salt concentrations, extreme temperatures (such as extreme drought or cold), high light intensity, and waterlogging.

3.5   What are the Indirect Mechanisms?

Indirect mechanisms employed by beneficial microorganisms include the control and prevention of plant diseases, which is an environmentally friendly approach. These microorganisms also act as biocontrol agents. Some of the major mechanisms of biocontrol activity include competition for soil-bound plant nutrients, niche exclusion, induced systemic resistance, and the production of antifungal metabolites.

Rhizobacteria produce various antifungal metabolites such as HCN, phenazines, pyrrolnitrin, 2,4-diacetylphloroglucinol, pyoluteorin, viscosinamide, and tensin. When interacting with plant roots, some rhizobacteria induce systemic resistance (ISR) in plants, which enhances their resistance against pathogenic bacteria, fungi, and viruses.

3.6   Quantifying the Efficiency of Microbial Fertilizers

The effectiveness of inoculated microbial fertilizers is influenced by various factors that need to be carefully addressed to enhance their efficiency in agriculture, particularly in fruit farming. One crucial factor is the quality of the product used, which is greatly influenced by the production process. Therefore, it is essential to focus on this important aspect in order to optimize the efficacy of microbial fertilizers.

3.7   The Process of Production of Microbial Fertilizer

The quality of a microbial fertilizer is determined by the production process employed. The population density of the mother culture and the quality of the final product are directly related (Stephens and Rask 2000). In the case of an inoculum consisting of multiple microorganisms, the interrelationships between these microorganisms and the plant's rhizosphere are of crucial importance. It is essential to understand the various modes of action of each microorganism involved, as their efficiency and potential antagonistic activities may overlap. Taxonomic diversity is not the sole determining factor for the relationship between microorganisms and the host plant; functional diversity plays a significant role (Maherali and Klironomos 2007).

The enhanced shelf life of the inoculant is crucial for ensuring the efficiency of a specific microbial fertilizer. Additionally, the biological traits of the inoculant are also important factors. Both of these aspects present challenges in determining the quality of a microbial fertilizer (Bashan et al. 2014). The formulation of a microbial fertilizer, whether it involves a single microorganism or a consortium, with or without additives, determines its efficacy. The formulation influences the shelf life during storage and transport, ensuring consistent persistence without compromising the quality or efficacy of the microbial fertilizers, thereby maximizing the benefits to the host plant postinoculation.

Various carriers are employed in the formulation process of microbial fertilizers, and they can affect the overall quality and efficacy. Granular inoculants have shown promising results, particularly under stressful soil conditions. Encapsulation of microbial strains into polymers allows for diverse composition and structure (Vassilev et al. 2005). However, there may be commercial limitations to the use of encapsulation in large-scale production (Bashan et al. 2014). The use of specific additives has the potential to improve the shelf life of microbial fertilizers (Bashan et al. 2014).

3.8   What are the Quality Standards for Microbial Fertilizers?

For effective and responsible marketing, it is essential for any manufactured product, including microbial fertilizers, to meet minimum specified quality standards. Insufficient scientific information provided on the label of the microbial fertilizer package can result in inconsistencies in field applications, ultimately affecting the market potential of the product. It is crucial to ensure a minimum quality standard that benefits consumers when they purchase a specific brand of microbial fertilizer. The distribution network also plays a role in influencing the quality of the microbial fertilizer. Numerous investigations have conclusively demonstrated that a reduction in the population of microbial inoculants, particularly under inadequate or improper storage conditions, leads to a decrease in the efficiency of the specific inoculant when applied in the field (Hartley et al. 2005).

3.9   The Persistence of the Microbial Fertilizer in the Soil

Due to the complex network of microorganisms present in the soil and rhizosphere, establishing the persistence and traceability of applied microbial strains in the soil is challenging (Torsvik and Ovreas 2002). Currently, there is no single technique available to accurately trace the persistence of a bioinoculant in the soil after application. Given the multitude of microbial fertilizers available in the market, monitoring the effectiveness of soil-applied bioinoculants becomes crucial for evaluating their efficacy and improving subsequent field application strategies. Therefore, there is an urgent need to standardize traceability techniques that rely on accurate and quantifiable measurements.

One approach for detecting released microorganisms in the environment is through PCR methods, which can identify both naturally occurring and externally introduced microbial strains (Stockinger et al. 2010). Reverse transcription polymerase chain reaction (RT-PCR) tests, commonly used for tracing the presence of lethal viruses like Ebola or COVID in the human body, can also be employed. By targeting specific genes of interest, these tests can provide information on the relative abundance of introduced strains of the microbial fertilizer within the microbial community, whether in the environment or soil. RT-PCR can be used to investigate the dynamics of the inoculated microbial fertilizer within the microbial community (Babiae et al. 2008).

3.10   The Role of the Host Plant

The introduced microbial population in the soil is subjected to significant pressure from the host plant, which ultimately impacts the plant's growth and yield through physiological and phenological effects. The host plant has the ability to modulate the release of compounds from its roots, leading to changes in the composition of rhizodeposits (Hartmann et al. 2009). These rhizodeposits vary over time, space, root position, and the plant's growth stage (Dennis et al. 2010). Therefore, selecting specific rhizosphere strains can be a more favorable approach (Marschner and Timonen 2005).

Root exudates contain compounds that can have both stimulatory and inhibitory effects on inoculated or rhizosphere microorganisms, influencing their ability to establish beneficial interactions with the plant (Hartmann et al. 2009). Microbial communities in the rhizosphere depend on other carbon pools present in the rhizosphere, such as microbial exudates, for their normal existence alongside the host plant (Dennis et al. 2010) and the introduced microbial strains. Root exudates play a crucial role in shaping the rhizosphere environment, particularly during the early stages of plant growth (e.g., very young seedlings) and the emergence of lateral roots. Consequently, applying microbial fertilizers to seeds and seedlings can enhance the effectiveness of the microbial fertilizer treatment.

3.11   Which are the Soil Conditions That Affect
the Microbial Action?

Both soil reaction (pH) and physical texture play a significant role in influencing the bacterial and fungal populations. Among these factors, soil pH is particularly critical as it determines the intensity of bacterial activity. Soils that are neutral in pH tend to have a higher microbial population density, while acidic soils can suppress microbial activity. In cases where the soil is acidic, it is essential to amend it before applying microbial fertilizers. This often involves the application of acidity-lowering substances like lime to raise the pH and create a more favorable environment for microbial growth and activity.

3.12   Competition from Inherent Microbial Population Against a Microbial Fertilizer

When an inoculant is introduced to the soil through a microbial fertilizer, it encounters competition from the existing microbial population. However, the exact nature of the interaction between these two populations is not yet fully understood. When host plants are inoculated with different microbial fertilizers, it can have a significant impact on the biological functions of the native microbial population. To gain a better understanding of these interactions, it would be valuable to identify genes that are involved in important enzymatic activities or in the interaction processes between the introduced microorganisms and the native microbial population. This knowledge could help in the development of microbial fertilizers that are specifically tailored for particular soils or crops, optimizing their effectiveness and promoting desired functions in the ecosystem.

3.13   How do Farmers' Practices Affect the Efficiency of Microbial Fertilizers?

3.13.1   Application of Fertilizers

The excessive use of synthetic fertilizers has a significant impact on the effectiveness of microbial fertilizers. Uncontrolled application of large quantities of synthetic fertilizers, such as excessive urea application in the pursuit of high crop yields (a characteristic of extractive farming known as the "green revolution"), can have detrimental effects on soil microflora and the performance of inoculated strains. Prolonged use of synthetic nitrogenous fertilizers can lead to a reduction in microbial activity, primarily due to the acidification of the soil, which is unfavorable for microbial growth, as mentioned earlier. Moreover, these fertilizers can have negative impacts on soil bacteria and vesicular-arbuscular mycorrhizal (VAM) communities (Toljander et al. 2008). The application of wastewater for irrigation purposes can also result in similar adverse effects on microbial activity. However, the use of consortia of phosphate-solubilizing bacteria (PSB) and potassium-solubilizing bacteria (KSB), along with the direct application of rock phosphate and potassium rocks, has shown promising results in increasing crop yield and enhancing the uptake of nitrogen (N), phosphorus (P), and potassium (K) in various vegetable crops grown in soils with inherent deficiencies of P and K (Supanjani et al. 2006).

3.14   What are the Management Practices in Soil That Affect the Efficiency of Microbial Inoculation?

Soil cultural practices, such as field preparatory plowing before crop sowing, inter-culture after sowing, and subsequent growth stages, have a significant impact on the soil microbial population. Studies have shown that fields where only organic matter was applied over a period of two decades exhibited enhanced and diverse bacterial and fungal populations compared to those treated with synthetic fertilizers (Berthrong et al. 2013). The effectiveness of vesicular-arbuscular mycorrhizal (VAM) inoculation can also be influenced by cultural practices, including pest management, combined application of synthetic and organic fertilizers, and irrigation regimes (Alguacil et al. 2014).

In arable soils, plowing practices can lead to a reduction in VAM fungal populations, such as Acaulospora, Gigaspora, Paraglomus, and Scutellospora, as these fungi are highly sensitive to tillage operations. Numerous studies have demonstrated that soil disturbances caused by plowing and nutrient status have adverse effects on fungal populations (Viti et al. 2010). Additionally, when considering substrate preparation for potted crop nurseries, the characteristics of the peat used can influence the colonization of roots by arbuscular mycorrhizal fungi (AMF) (Ma et al. 2007). These findings highlight the importance of considering the specific conditions, especially soil conditions, where microbial fertilizer strains will be used when developing them.

3.15   Application Method of the Inoculant

The effectiveness of applied microbial fertilizers is influenced by the method of application. Although there is limited research in this area (Bashan et al. 2014), granulated forms of microbial fertilizers can be easily applied using conventional equipment. In horticultural crops, specific equipment has been developed for the distribution of microbial fertilizers on a small scale, which is particularly beneficial in fruit farming (Wawrzynczak et al. 2011).

However, when liquid microbial fertilizers are applied using a regular sprayer, it has been observed to have an impact on microbial viability. Continued application can result in a decrease of viable cell count by up to 50%. The volume of water used and the addition of adjuvants also influence the delivery and efficacy of spores (Bailey et al. 2007). In some cases, foliar application of microbial fertilizers can be considered, especially for endophytic species that enhance the growth and yield of fruit crops. The effectiveness of such applications has been demonstrated in various fruit farming scenarios (Pirlak et al. 2007).

3.16   What are the Environmental Factors That Affect Microbial Fertilizers' Efficiency?

The following factors greatly influence the efficacy of the microbial fertilizers:

1. Effect of temperature

Temperature plays a crucial role in the shelf life and effectiveness of microbial fertilizers. The optimal temperature varies depending on the specific strain. During the cropping season, microbial colonization occurs optimally at field temperature (soil temperature). Liquid formulations of microbial fertilizers typically grow best at around 37 °C ± 1 °C and can tolerate temperatures up to 45 °C ± 1 °C for about 1 year after soil inoculation. However, solid formulations have a shorter shelf life of approximately 6 months, and their effectiveness rapidly deteriorates as the temperature rises to 35 °C (Patel 2012). Therefore, it is important to maintain ideal soil/ambient temperature during storage and field application.

2. Effect of environmental acclimatization

Variations in environmental conditions can negatively impact microbial efficiency. While the efficiency of liquid formulations of N and P microbial fertilizers remains relatively consistent across different environmental situations, carrier-based formulations are more affected. Factors such as temperature, humidity, leaf surface, root exudates, and competing microbial populations can deactivate the applied microbial strain. Physical loss of the inoculated strain can also occur through wind, rain, or surface leaching during heavy rainfall (edaphic factors).

3. Effect of moisture

Moisture and humidity are crucial factors that affect the storage, stability, and activity of microbial fertilizers. While some microbes require moisture for their activity, carrier-based inoculants can experience stress due to dry carrier materials during transport and storage. Microbial strains need moist conditions for establishment and subsequent activity, which are provided by liquid formulations containing humectants. The direct effect of humidity on spore-forming bacteria in liquid forms is relatively low.

4. Effect of sunlight intensity

Sunlight intensity directly affects microbial activity due to its impact on temperature fluctuations. Microorganisms are directly and adversely affected by the harmful effects of ultraviolet (UV) rays. Chemicals can be added to microbial fertilizers as sunscreens to reflect, scatter, or convert harmful UV rays to less harmful wavelengths. Such sunscreens are available in carrier-based formulations to counter the adverse effects of sunlight. It is important to select or screen microbial fertilizers that can tolerate high-intensity sunlight and resulting high temperatures.

5. Effect of the reaction (pH) of the microbial fertilizer

The reaction (pH) is particularly important for liquid microbial fertilizers compared to solid formulations. Microbial populations can be inactivated at very high or very low pH levels. Adjustments can be made by adding sodium hydroxide (NaOH) to increase pH or hydrochloric acid (HCl) to decrease pH, ensuring a longer shelf life for microbial fertilizers.

6. Effect of the organic matter or organic carbon content in the soil

Soil organic matter, derived from sources such as crop residues and animal manure, contains organic carbon that directly interacts with the microbial flora in microbial fertilizers. Organic carbon serves as the primary energy source for specific bacteria carried by microbial fertilizers. Therefore, the efficiency of microbial fertilizers is influenced by the organic carbon content in the soil, whether native or added through organic matter.

How are Microbial Fertilizers Formulated? What is a Multi-strain Inoculant?

The efficiency of microbial fertilizers is determined by the following two most important factors:

1. The dosage of the specific inoculum.
2. The method of application of the inoculum.

3.17 How is the Seed Treated?

The inoculation process consists of the following three steps:

Step 1: Seed treatment

To treat the seeds, a 200 g package of the specific inoculant is mixed with 200 ml of rice gruel or a 40% gum Arabic solution to create a slurry. The seeds required to plant 1 hectare/acre are then thoroughly mixed in the slurry to ensure a uniform coating of the inoculant on the entire seed lot. Afterward, the inoculated seeds are dried on a polythene sheet under shade for half an hour in a clean environment to avoid any external contamination. The dried seeds should be sown in the designated field within 24 h. Typically, one 200 g packet of the inoculant is sufficient to treat 10 kg of seed material.

Step 2: Dipping of the seedlings

For transplanted crops like wetland rice, two packets of the inoculant (2 x 200 g) are mixed with 2 liters of double-distilled water to ensure sterility. The seedlings required for planting one acre are then dipped in the slurry for half an hour. Following this step, the seedlings are ready for transplantation in the main rice field.

Step 3: How to apply the seed–inoculant mixture in the main field?

In this step, four packets of the inoculant (4 x 200 g) are mixed with 20 kg of dried and finely powdered farmyard manure (FYM) or farm compost. The resulting mixture is then broadcast evenly over 1 acre of the main field. The amount of inoculant to be applied per plant depends on the duration of the specific crop. For short-duration crops (approximately 6 months), 10–25 g per plant is used. For long-duration crops, more than 6 months, 25–50 g per plant is applied during the first year, and from the second year onward, 50 g per plant is used.

3.18 What are the Specific Recommendations for Fruit Crops?

The following recommendations may be noted in case of fruit and other edible crops, as per data given in Table 3.1.

3.19 Which are the Most Important Precautions to be Taken When Using Microbial Fertilizers?

The following points should be strictly adhered to while using microbial fertilizers:

Point No. 1: Ensure that all materials used in the production of microbial fertilizers are within their expiration date.
Point No. 2: Store the package containing the inoculant in a well-ventilated, cool, and shaded place until it is ready to be used.
Point No. 3: Use the microbial fertilizer only for the specified crop mentioned on the package. This is especially important when using Rhizobium culture.
Point No. 4: Open the package containing the microbial fertilizer just before field use, and immediately sow the inoculated seeds after mixing.

Table 3.1 Details of the fruit/other edible crops and the inoculant rate used

Fruit/other edible crops	Rate of inoculant used
Tomato, brinjal, chili, cauliflower, cabbage	1 kg *Azotobacter* + 1 kg PSB for and other transplanted crops, like rice 1 acre as seedling root dip method
Potato, ginger, Colocasia, turmeric	*Azotobacter* or *Azospirillum* 4 kg + 4 kg PSB for 1 acre mixed with 100–200 kg of farm compost and applied in the field
Tea, coffee, mulberry, rubber, other fruit trees	*Azotobacter* or *Azospirillum* 2–3 kg + PSB 2–3 kg mixed with 200 kg farm compost for 1 acre and applied in the field

Note: The inoculum rate varies with the seed rate, crop, and soil edaphic factors, which need to be standardized before the field application is made

Point No. 5: Avoid using warm or hot water when preparing or using microbial fertilizers.

Point No. 6: When treating seeds with both a fungicide and an insecticide, apply the fungicide first, followed by the insecticide, and finally, use the microbial fertilizer.

Point No. 7: If the microbial fertilizer is to be used in highly acidic or saline-alkaline soil, make appropriate soil amendments to bring the soil pH close to neutrality (around pH 7.0) before sowing the seeds with the specific microbial fertilizer.

Point No. 8: Prior to using the microbial fertilizer, ensure that the field soil is adequately fertilized with optimal nutrients in sufficient quantities. Conduct a reliable soil test beforehand to determine the optimal fertilizer application. This step is particularly important when using Rhizobium as an inoculant, as it helps promote good microbial activity.

3.20   What is a Consortium of Microbial Fertilizers?

A plant's/crop's growth and performance can be directly influenced by the application of microbial fertilizers. These fertilizers enhance the bioavailability of soil nutrients and promote the production of plant hormones, resulting in positive effects on plant/crop health. Moreover, microbial fertilizers contribute to effective soil pathogen management and help reduce the reliance on agrochemicals, making them an environmentally viable option for modern agriculture (Aloo et al. 2019). As opposed to extractive farming, which poses significant environmental risks, bio farming represents a better and more sustainable choice for the twenty-first century.

Soil microbes typically exist as complex communities rather than single cells, found in the vicinity of plant root canopies, bulk soils, endophytes, or even on plant surfaces (Lorito et al. 2006). These microflora form part of a diverse family of microbiota that adapt to environmental stresses through beneficial interactions within the microbial community (Hacquard et al. 2015). The functional diversity and succession of different microbial components within a microbial consortium (MIC) contribute to the physiological functions of ecosystems. The microbial balance in a specific MIC involves a dynamic interplay between actively growing cells and nondividing cells. Microbial populations communicate with each other through a phenomenon called "quorum sensing," exchanging specific chemical signals that illustrate the remarkable intricacies of nature. Natural microbial consortia possess properties like stability, functional diversity, and the ability to perform complex functions, generating interest in the development of synthetic consortia for application (Minty et al. 2013).

Traditionally, microbial inoculants have focused on single specific strains. However, recent research has shifted toward the development of synthetic multi-strain inoculants, whether bacterial or fungal, based on the premise that a combination of multiple strains would be more effective and perform better than a single strain (Vorholt et al. 2017). While single-strain microbial fertilizers can be effective, mixed consortia of microbial fertilizers have the advantage of acclimatizing to a wide range of environmental conditions and exhibiting multifunctional activities (Sarma et al. 2015). Synthetic mixtures of bacterial/fungal strains have been investigated as microbial fertilizers or antagonistic agents, demonstrating plant growth-promoting (PGP) activities and plant-pathogen suppressive abilities (PPSA). However, it is important to consider potential antagonistic interactions among the microorganisms in the mixture, as these interactions may reduce the expected positive effects (Sarma et al. 2015). Compatibility among different microbial strains in a mixture remains an unresolved issue in the development of effective multi-strain microbial consortia as inoculants (Sarma et al. 2015).

In summary, this book emphasizes that bio farming is a more appropriate and sustainable alternative to the extractive farming practices of the past. By harnessing the potential of microbial fertilizers and promoting the use of diverse microbial communities, agriculture can be transformed into a more environmentally friendly and productive system.

References

Alguacil MM, Torrecillas E, Lozano Z, Torres MP, Roldan A (2014) *Prunus persica* crop management differentially promotes arbuscular mycorrhizal fungi diversity in a tropical agro-ecosystem. PloS ONE 9(2):e88454. https://doi.org/10.1371/journal.pone.088454

Aloo BN, Makumba BA, Mbega ER (2019) The potential of bacilli rhizobacteria for sustainable crop production and environmental sustainability. Microbiol Res 219:26–39

Babiae KH, Schauss K, Hai B (2008) Influence of different *Sinorhizobium meliloti* inocula on abundance of genes involved in nitrogen transformations in the rhizosphere of alfalfa (*Medicago sativa* L). Environ Microbiol 10(11):2022–2930

Bailey KL, Carisse O, Leggett M, Holloway G, Leggett F, Wolf TM, Shivpuri A, Derby JA, Caldwell B, Geissler HJ (2007) Effect of spraying adjuvants with the biocontrol fungus *Microsphaeropsis ochracea* at different water volumes on the colonization of apple leaves. Biocontrol Sci Tech 17:1021–1036

Bashan Y, De-Bashan LE, Prabhu SR, Hernandez JP (2014) Advances in plant growth – promoting bacterial inoculant technology: formulations and practical perspectives (1998-2013). Plant Soil 378:1–33

Berthrong ST, Buckley DH, Drinkwater LE (2013) Agricultural management and labile carbon additions affect soil microbial community structure and interact with carbon and nitrogen cycling. Microb Ecol 66:158–170

Dennis PG, Miller AJ, Hirsch PR (2010) Are root exudates more important than other sources of rhizodeposits in structuring rhizosphere bacterial communities? FEMS Microbiol Ecol 72:313–327

Hacquard S, Garrido-Oter R, Gonzalez A, Spaepen S, Ackermann G, Lebeis S, McHardy AC, Dangl JL, Knight R, Ley R, Schuze-Lefert P (2015) Microbiota and host nutrition across plant and animal kingdoms. Cell Host Microbe 17(5):603–616

Hartley EJ, Germell LG, Slattery JF, Howieson JG, Herridge DF (2005) Age of peat-based lupin and chickpea inoculants in relation to quality and efficacy. Anim Prod Sci 45:183–188

Hartmann A, Schmid M, Van-Tuinen D, Berg G (2009) Plant-driven selection of microbes. Plant Soil 321:235–257

Lorito M, Woo S, Iaccarino M, Scala F (2006) Antagonistic microorganisms. In: Laccarino M (ed) Microorganims beneficial to plants. Idelson-Gnocchi Publ, Napoli, pp 145–175

Ma N, Yokoyama K, Marumoto T (2007) Effect of peat on mycorrhizal colonization and effectiveness of the arbuscular mycorrhizal fungus *Gigaspora margarita*. Soil Sci Plant Nutr 53:744–752

Maherali H, Klironomos JN (2007) Influence of phylogeny on fungal community assembly and ecosystem functioning. Science 316:1746–1748

Marschner P, Timonen S (2005) Interactions between plant species and mycorrhizal colonization of the bacterial community composition in the rhizosphere. Appl Soil Ecol 28:23–36

Minty JJ, Singer ME, Scholz SA, Bae C, Ahn J, Foster CE, Liao JC, Lin XN (2013) Design and characterization of synthetic fungal-bacterial consortia for direct production of isobutanol from cellulosic biomass. Proc Natl Acad Sci 110(36):14592–14597

Patel BC (2012) Advanced method of preparation of bacterial formulation using potash mobilizing bacteria that mobilize potash and make it available to crop plant. Patent No. WIPOWO/2011/154961

Pirlak L, Turan M, Sahin F, Esitken A (2007) Floral and foliar application of plant growth promoting rhizobacteria (PGPR) to apples increases yield, growth, and nutrient element contents of leaves. J Sustain Agric 30(4):145–155

Sarma BK, Yadav SK, Singh S, Singh HB (2015) Microbial consortium-mediated plant defense against phytopathogens: readdressing for enhancing efficacy. Soil Biol Biochem 87:25–33

Stephens JHG, Rask HM (2000) Inoculant production and formulation. Field Crops Res 65:249–258

Stockinger H, Kruger M, Schubler A (2010) DNA barcoding of arbuscular mycorrhizal fungi. New Phytol 187:461–474

Supanjani, Han HS, Jung SJ, Lee KD (2006) Rock phosphate potassium and rock solubilizing bacteria as alternative sustainable fertilizers. Agron Sustain Dev 26:233–240

Toljander JF, Santos-Gonzalez JC, Tehler A, Finlay RD (2008) Community analysis of arbuscular mycorrhizal fungi and bacteria in the maize mycorrhizosphere in a long-term fertilization trial. FEMS Microbiol Ecol 65:323–338

Torsvik V, Ovreas L (2002) Microbial diversity and function in soil: from genes to ecosystems. Curr Opin Microbiol 5:240–245

Vassilev N, Nikolaeva I, Vassileva M (2005) Polymer based preparation of soil inoculants: applications to arbuscular mycorrhizal fungi. Rev Environ Sci Biotechnol 4:235–243

Viti C, Tatti E, Decorosi F, Lista E, Rea E, Tullio M, Sparvoli E, Giovannetti L (2010) Compost effect on plant growth – promoting rhizobacteria and mycorrhizal fungi population in maize cultivation. Compos Sci Utiliz 18(4):273–281

Vorholt JA, Vgel C, Carlstrom CI, Mueller DB (2017) Establishing causality: opportunities of synthetic communities for plant microbiome research. Cell Host Microbe 22(2):142–155

Wawrzynczak P, Bia3 (3 superscript please)kowski P, Rabcewicz J, Plaskota M, Gotowicki B (2011) Application of biofertilizers and biostimulants in organic orchards. In: Sas-Paszt L, Malusa E (eds) Proceedings of Ogolnopolsk Naukow 1 (1 superscript please) Konferencje Ekolgieczn1 (1 superscript please), Osi1(1 superscriptplese)gniecia imoliwooeci rozwoju badan iwdro en wekologicznej produkcji ogrodniczej. Skierniewice 6-7/10/2011 Skierniewice. Institute of Horticultural Research, pp 85–86

Printed in the United States
by Baker & Taylor Publisher Services